QUANTITATIVE
MASS SPECTROMETRY

QUANTITATIVE
MASS SPECTROMETRY

BRIAN J. MILLARD

The Institute of Neurology,
London

LONDON · PHILADELPHIA · RHEINE

PHYSICS

Heyden & Son Ltd., Spectrum House, Hillview Gardens, London NW4 2JQ.
Heyden & Son Inc., 247 South 41st Street, Philadelphia, P.A. 19104, U.S.A.
Heyden & Son GmbH, Münsterstrasse 22, 4440 Rheine/Westf., Germany.

BIOLOGY

ISBN 0 85501 156 4

Printed in Great Britain by Galliard (Printers) Ltd., Great Yarmouth, England

CONTENTS

PREFACE

The driving force for the rapid expansion of mass spectrometry in the 1960s was the recognition by organic chemists of the vast amount of structural information which could be obtained by the technique. Workers in the area of medicine and biochemistry soon became convinced of the value of mass spectrometry as a qualitative tool, but naturally, having identified compounds of biological importance in their extracts, their attention turned to the quantitative determination of these substances. The recent growth of mass spectrometry as a quantitative technique is a direct result of the demands of these workers, and indeed the pioneering work in mass spectrometry in this area was carried out by biochemists and pharmacologists. Thus, not surprisingly, the majority of the examples in this book concern the determination of compounds of biochemical importance.

Although the increasing importance of quantitative mass spectrometry in the last few years has been obvious to anyone attending conferences on mass spectrometry, particularly where applied in medicine and biochemistry, this importance has been recognized in print only by the publication of a few reviews of specific areas. This book has been written to emphasize the fact that quantitative mass spectrometry is now a subject in its own right; many vastly different problems from a variety of disciplines can be tackled by a broadly similar approach. From this point of view, the problems of the perfume chemist are not very different from those of the pharmacologist or the quality control chemist. Each wishes to quantify a very small amount of a substance in a very complex mixture, and this book is intended to help each of them.

The book is also meant to bring to the attention of those who have not yet considered mass spectrometry to be a quantitative technique the considerable progress which has been made in harnessing the unrivalled sensitivity and specificity to quantitative determination. If such workers in other areas of research can be encouraged to apply mass spectrometry to their problems, then their experience gained in other analytical techniques cannot but help to ensure the continuing progress of quantitative mass spectrometry.

Finally, I would like to express my sincere thanks to Professors Catherine Fenselau and Clyde Williams for devoting so much of their time to reading the typescript, and for their many helpful suggestions which have been incorporated into the final book.

London   Brian J. Millard
September, 1977

INTRODUCTION

There are still a few adherents to the school of thought that analytical instruments can be treated as 'black boxes', into which solutions are fed, and from which numerical results are obtained without worrying too much about what occurs inside the box. There are many occasions where such an approach can be reasonably satisfactory, for example when the black box is a simple, rugged instrument with few knobs to turn, and when the particular analysis is not pressing the capabilities of the instrument near to its limits of performance. By no stretch of the imagination can mass spectrometry be considered to be amenable to such an approach. Mass spectrometry is a relatively expensive technique, and more often than not analysts will have turned to it as a last resort because less expensive instrumentation is either not sensitive enough or not specific enough. In such circumstances the mass spectrometer will frequently be operating near to its limits.

If the instrument is not well maintained its sensitivity will inevitably decrease, sometimes by a factor of many hundreds, and the precision of the quantitative analyses will fall off, perhaps to intolerable levels. A fundamental understanding of how ions are produced, separated and detected is essential to the efficient utilization of a mass spectrometer, and the first two chapters of this book are devoted to this aspect, with a discussion of basic instrumentation and the more recent developments which are relevant to quantitative work.

The error involved in a measurement is a prime consideration in analytical work, and the various sources of error, from the initial sample manipulation to the final measurement of a response on the chart paper or by a computer, are covered in Chapter 3. Any discussion of errors inevitably moves into the realm of statistics, and simple statistical operations are utilized throughout the text. It was considered useful to gather together the more important equations and tables and place them in an Appendix at the end of this book for rapid reference.

The construction of calibration curves is a much larger topic than is apparent from the literature of quantitative mass spectrometry, and much more thought should be put into their determination. The various points which should be considered in connection with calibration lines are covered in Chapter 4.

Although, when quantitative mass spectrometry is mentioned, one's thoughts turn almost automatically to the use of selected ion monitoring in a combined gas chromatography mass spectrometry system, a study of the literature shows that much useful work is carried out by means of direct sample introduction. Indeed, there are many compounds that are too involatile or thermally unstable to pass through a gas chromatograph, and direct introduction is the only method of obtaining the necessary data. Without a doubt, this is a much neglected area, and one which would benefit considerably from further applications such as the determination of substances in biological tissues under high mass spectrometric resolving power.

At the moment, selected ion monitoring in a g.c.m.s. system is the fastest growing area of quantitative mass spectrometry, and therefore, not surprisingly, accounts for the largest chapter in this book. The three most important aspects of the method to consider are sensitivity, specificity, and the choice of internal standard. The first two are, to some extent, capable of being traded off against each other. For small amounts of a given component, single ion monitoring will be the most sensitive technique, but may result in a loss of specificity when compared with the monitoring of say four different ions in the spectrum. A choice between 1, 2, 3 or 4 channel monitoring will depend therefore very much on the nature of the problem, such as how little of the compound of interest is present, and how many interfering substances are present.

The use of single ion monitoring, which generally means that homologous internal standards have to be used which yield an ion in common with the compound of interest, has perhaps been emphasized rather strongly in this book. This is mainly an attempt to achieve a balance vis-à-vis the current literature on internal standards, where stable isotope labelled analogues are usually preferred. There are many instances where stable isotope labelled analogues are preferable, but it can be argued that there are many more occasions where they are chosen because they are fashionable, rather than because the relative merits of various internal standards have been tested rigorously and the labelled compounds have been shown to be superior.

It is intended that this book should stimulate a more questioning approach to the whole concept of quantitative determinations by mass spectrometry, so that the development of an assay is given the careful thought and attention to detail necessary if disappointment is to be avoided.

CHAPTER 1

BASIC INSTRUMENTATION

A mass spectrometer may be defined for our purposes as an instrument that produces ions and then separates them according to their mass-to-charge (m/e) ratio. The ions can be produced in many ways; for example, by electron impact, chemical ionization, charge transfer, field ionization, field desorption and in a spark source. The ions can also be separated in many ways, by magnetic, quadrupole and time-of-flight analysers. Both positive and negative ions are formed in the mass spectrometer source, but the majority of instruments are designed for the efficient formation and examination of positive ions.

ION PRODUCTION
Electron impact

In an electron impact (e.i.) source, the sample is bombarded with a beam of electrons in the vapour phase. Usually, the energy of this beam can be adjusted from about 5 to 80 eV, but by convention operates normally at 70 eV. Unless otherwise stated, it can be assumed that published mass spectra have been obtained at this beam energy. It has been estimated that even at this high energy in the most efficient sources available, only about one molecule in a thousand is ionized. Of all the ions formed, only a small proportion are negatively charged. In the case of positive ions, most are singly charged, a few are doubly charged, and very occasionally triply charged ions may be formed. The occurrence of doubly charged ions is very dependent on the type of compound being ionized, and such ions are more prevalent in the mass spectra of aromatic compounds containing several rings.

For an organic compound M, the situation is as follows:

$$M + e \rightarrow [M]^{\overset{+}{\cdot}} + 2e$$
$$M + e \rightarrow [M]^{2+} + 3e$$
$$M + e \rightarrow [M]^{3\overset{+}{\cdot}} + 4e$$

The $+$ notation for the species [M]$^+$ implies that the latter is an ion radical, i.e. an ion with an odd number of electrons. The doubly charged species has an even number of electrons. Ions derived from the molecular ion by fragmentation may be even or odd electron species, depending on the identity of the precursor ion and the nature of the fragment lost.

The amount of energy necessary to remove an electron from M is called the ionization potential (IP). Ionization potentials can be measured by the mass spectrometer and by other techniques. The energy needed to form a particular fragment ion can also be measured, and this is called the appearance potential (AP) of the ion in question. When the electron beam is operating at 70 eV, an average of several electron volts of energy in excess of the ionization potential is transferred to the molecular ion. It is this excess energy which causes the molecular ion to fragment. A hypothetical energy distribution in the molecular ions formed from a compound under electron impact may look like the example shown in Fig. 1.1.

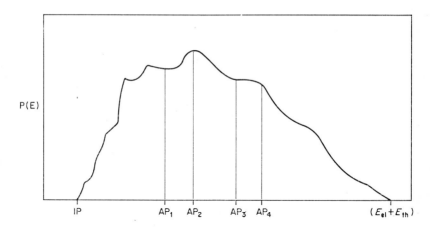

Fig. 1.1. A hypothetical energy distribution in molecular ions formed from an organic compound under electron impact. $P(E)$ represents the probability of occurrence of an ion with energy E. IP is the ionization potential of the compound. AP$_1$ to AP$_4$ are the appearance potentials of four different fragment ions in the mass spectrum. $(E_{el} + E_{th})$ is the sum of the electron beam energy and the thermal energy of the molecule.

The appearance potentials AP$_1$ to AP$_4$ represent the energy required to form fragment ions F$_1$ to F$_4$, where F$_1$ is the most easily formed of these ions, i.e. requires least energy. The fragment ions may be ions [F$_3$]$^+$ or ion radicals [F$_1$]$^+$ and may be formed from two or more precursor ions.

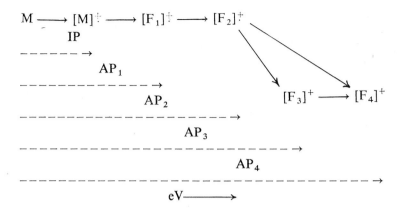

Ions are considered to be more stable than ion radicals, and tend to fragment by loss of a neutral molecule to form another ion, rather than losing a radical to form an ion radical. Thus, fragmentations involving the consecutive loss of two radicals are rare.

Taking a simplified view, only those molecular ions with energies between IP and AP_1 will be observed as molecular ions in the mass spectrum. Ions with energy greater than AP_1 will have decomposed to the particular fragment ion for which they have sufficient energy. Therefore, the abundance of the molecular ion in the final mass spectrum will depend upon what proportion of the total area under the curve is represented by the area between IP and AP. The maximum energy which an ion may have, i.e. at the point represented by the high limit of the curve, will be the sum of the thermal energy of the molecule E_{th} and the energy of the electron beam E_{el}. For a molecule such as 1,2-diphenylethane, the average thermal energy has been calculated as 0.3 eV at 75 °C and 0.7 eV at 200 °C.[1] This is significant when compared with ionization potentials which are typically in the range 7–20 eV.

The simple view outlined above neglects what is called the kinetic shift. In a magnetic instrument, normal fragment ions observed in the mass spectrum are those which have been formed in the source. Since the residence time of ions in the source is about 10^{-6} s, the ions have to be formed with a unimolecular rate constant greater than 10^6 s^{-1} in order to be observable in the mass spectrum. The kinetic shift is the excess energy required above the theoretical appearance potential in order for decomposition to occur with this rate constant. Thus, the appearance potentials AP_1 to AP_4, as determined in the mass spectrometer, will be higher than the theoretical values.

A typical e.i. mass spectrum, for example that shown for chlorpromazine in Fig. 1.2, exhibits a large number of fragment ions in addition to the molecular ion.

In some cases, for example barbiturates, the molecular ion decomposes so readily that it is not observed in the spectrum. Since the most important item of information to be gleaned from a mass spectrum is the molecular weight of the

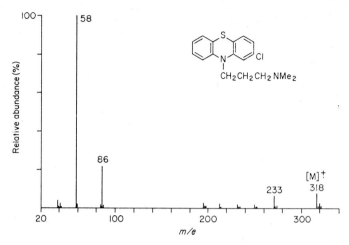

Fig. 1.2.   Mass spectrum of chlorpromazine. The molecular ion is at m/e 318.

compound, the major disadvantage of using e.i. conditions on an unknown compound is that it is never quite clear that the highest m/e value encountered is that
of the molecular ion. One way of surmounting this problem is to use the technique of metastable refocusing,‡ whereby it is possible to determine the precursor
ions for those ions which are of interest. Thus, it can be determined whether or
not the supposed molecular ion has a precursor, and if so, its m/e value may be
ascertained. At present this method is not easily applicable to samples eluting
from a gas chromatograph into the mass spectrometer source, and therefore is
of limited interest to, say, workers in the biomedical field. The most useful
solution to the problem is to use the softer ionization techniques of chemical
ionization, field desorption or fission fragmentation. These techniques can reduce
the wide energy spread of molecular ions, as shown in Fig. 1.1, to a very narrow
band, so that insufficient energy is transferred to cause any appreciable fragmentation. These alternative ionization methods must always be considered to
be complementary to electron impact ionization, mainly because no structural
information is given simply from knowledge of the molecular weight of an
unknown compound. In general, the more fragment ions there are in a mass
spectrum, the greater is the amount of structural information that may be
obtained.

A much larger proportion of the total ion current will be carried by the
$[M + 1]^+$ ions typical of chemical ionization spectra than is carried by the
molecular or fragment ion in the e.i. spectrum. If it is allowed that the total yield
of ions from a given weight of sample is approximately the same, this means that
there is a greater sensitivity for the particular compound. Another factor which
becomes especially important when working with complex mixtures is that the

‡ See Chapter 2

total number of different ions produced from this large number of compounds is greatly reduced, thereby lessening the chances that some other compound can interfere with the analysis by producing an ion of the same m/e value as the one to be monitored.

Once a compound of interest has been identified and attention turns towards its quantitative determination, in the majority of cases it can be argued that chemical ionization in particular is of greater use than electron impact ionization.

Chemical ionization

In chemical ionization (c.i.) mass spectrometry, the compound under investigation is ionized by reaction with a set of reagent ions. These reagent ions are formed from the reactant gas by a combination of electron impact ionization and ion–molecule collisions. The proportion of compound to reactant gas is usually of the order of 1 to 1000, so that electron impact ionization of the compound does not occur. One of the most popular reactant gases is methane, and the electron impact ionization and ion–molecule reactions of methane can be summarized as follows:

$$CH_4 \rightarrow [CH_4]^{+\cdot}$$
$$[CH_4]^{+\cdot} + CH_4 \rightarrow [C_2H_5]^+ + CH_3 \cdot$$
$$[CH_3]^+ + CH_4 \rightarrow [C_2H_5]^+ + H_2$$
$$[CH_3]^+ + 2CH_4 \rightarrow [C_3H_7]^+ + 2H_2$$
$$[CH_2]^{+\cdot} + 2CH_4 \rightarrow [C_3H_5]^+ + 2H_2 + H \cdot$$
$$[CH_2]^{+\cdot} + CH_4 \rightarrow [C_2H_4]^{+\cdot} + H_2$$
$$[CH_2]^{+\cdot} + CH_4 \rightarrow [C_2H_3]^+ + H_2 + H \cdot$$

At a source pressure of about 1 Torr, the ions formed from methane consist mainly of $[CH_5]^+$ ions (48%), with lesser amounts of $[C_2H_5]^+$ (40%) and

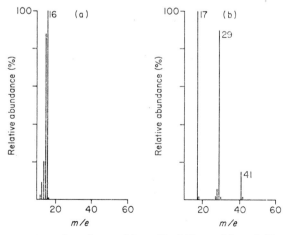

Fig. 1.3.   Mass spectra of methane: (a) at 10^{-5} Torr; (b) at 1 Torr. (Reproduced with permission from Ref. 2.)

$[C_3H_5]^+$ (6%).[2] This can be seen in Fig. 1.3, where the mass spectrum of methane is shown for normal source pressures and at the high pressure of 1 Torr.

The reactant ions $[C_2H_5]^+$ and $[CH_5]^+$ react with the sample (BH) mainly by proton transfer or hydride abstraction:

$$[CH_5]^+ + BH \rightarrow [BH_2]^+ + CH_4$$
$$[C_2H_5]^+ + BH \rightarrow [BH_2]^+ + C_2H_4$$
$$[C_2H_5]^+ + BH \rightarrow [B]^+ + C_2H_6$$

To a lesser extent alkyl transfer reactions can occur to give $[M + 29]^+$ and $[M + 41]^+$ ions. Thus,

$$BH + [C_2H_5]^+ \rightarrow [BHC_2H_5]^+$$
$$BH + [C_3H_5]^+ \rightarrow [BHC_3H_5]^+$$

The ions $[M + 1]^+$ $([BH_2]^+)$ or $[M - 1]^+$ $([B]^+)$ have often been described as *quasimolecular* ions, but the term is now falling into disuse. Since these ions are even electron species, they are inherently more stable than the $[M]^{+}$ ion produced

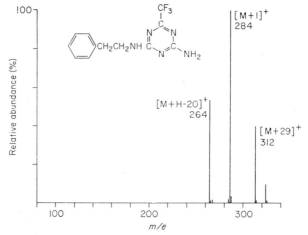

Fig. 1.4.   Methane c.i. mass spectrum of a derivative of phenformin. (Reproduced with permission from S. B. Matin, J. B. Knight, J. H. Karam and P. H. Forsham, *Biomed. Mass Spectrom.* **1**, 320 (1974).)

by electron impact. Furthermore, the amount of energy transferred to the quasi-molecular ion is much lower than that transferred in an e.i. source, but is highly dependent upon the reagent gas used. Because of these factors, the amount of fragmentation is usually greatly reduced, and the quasimolecular ion is normally the most intense ion in the mass spectrum. This is illustrated in Fig. 1.4, where the methane c.i. spectrum of a derivative of phenformin is given. The only significant fragmentation is the formation of an $[M + H - 20]^+$ ion. The alkyl transfer ion is intense.

There are many cases where the quasimolecular ion is not the most intense ion in the spectrum, however, for example the prostaglandins,[3] where the $[M + 1]^+$ ion loses one and two molecules of water to give more intense fragment ions.

Although earlier work on c.i. mass spectrometry was carried out using methane as reagent gas, much effort is being directed at present to the selection of reagent gases which will control the amount of fragmentation and also selectively ionize particular components in a mixture. Isobutane is a very popular reagent gas, where at high pressures the $[t\text{-}C_4H_9]^+$ ion accounts for over 90% of the ions formed. This ion is a weaker Brønsted acid than the $[CH_5]^+$ ion and transfers a proton with much less energy. Consequently, isobutane c.i. spectra are characterized by much less fragmentation, as shown in a study of a large number of drugs by Milne et al.[4] With the exception of aspirin, only $[MH]^+$ ions were produced. Isobutane transfers a proton only to the more basic compounds, so that strong acids are not ionized well. Frequently $[C_4H_9]^+$ is added to compounds such as alcohols.

Argon water mixtures have also been used successfully in c.i. work.[5] In a study of explosives, it was found that while methane and isobutane gave no $[MH]^+$ ions, water was successful in yielding intense $[M + 1]^+$ and $[M + 19]^+$ ions.[6] Ammonia has also been used, for example in a study of mono- and disaccharides,[7] where molecular weight information was obtained. Nitric oxide can condense with ketones, esters and carboxylic acids to give $[M + NO]^+$ ions and can also abstract hydride from aldehydes and ethers.[8]

There are some slight disadvantages to c.i. mass spectrometry which should be mentioned. In most cases the spectra are extremely temperature dependent, leading to poor reproducibility, a very important factor in quantitative work. The tight source design, combined with the fact that c.i. mass spectrometry is a high pressure technique, means that the source has to be cleaned much more frequently than is the case with e.i. sources. It is also extremely important to use high purity reagent gases, since the impurities present can lead to a very high background spectrum. Isobutane c.i. spectra frequently show cluster ions due to reaction with water present, so that very dry gases should be used. In combined gas chromatograph mass spectrometer (g.c.m.s.) systems, current practice is not to use the reagent gas as the carrier gas (the number of useful carrier gases is limited anyway), but to bleed the reagent gas into the system at a point after the molecular separator, since much better control over pressure, for example, is maintained. There are references, however, to the use of methane as a combined carrier gas and c.i. reagent gas.[8–10]

Three reviews by Field[11–13] are to be recommended as introductory reading to the topic of chemical ionization.

Charge exchange

In the above account, the reagent gases were shown to function mainly by transferring H^+ with the consequent formation of even electron ions. There is another category of c.i. reactions in which the reagent ions transfer an electron to form an odd electron species. This is called charge exchange or charge transfer, and usually involves a rare gas such as helium, argon or xenon. Since an odd molecular ion is formed, not surprisingly, charge exchange spectra look fairly similar to e.i. spectra, although the degree of fragmentation can be controlled. It is this latter point that is important, for the molecular ion is much enhanced compared with e.i. spectra, but one still has the advantage of structurally significant fragment ions. A mixture of nitric oxide (5–10%) in helium, nitrogen or argon has proved to be especially useful for enhancing the molecular ion, for example in the case of the trimethylsilyl (TMS) ethers of bile acid methyl esters.[14,15] Figure 1.5 shows a charge exchange mass spectrum of heroin obtained through

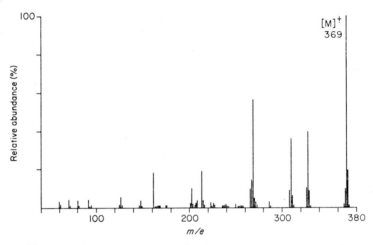

Fig. 1.5.   Charge exchange spectrum of heroin via a g.c.m.s. system. Reagent gas 10% NO in N_2. (Reproduced with permission from Ref. 16.)

a g.c.m.s. system, using 10% nitric oxide in nitrogen.[16] One percent mixtures of nitric oxide in nitrogen have also been used.[17]

Field ionization

The technique of field ionization is well described in an early book by Beckey, the pioneer of the technique.[18] The anodes in early field ionization (f.i.) sources were sharp blades, wires or points. Present-day commercial sources employ wire emitters, and cathodes with slits in them. A high voltage of the order of 2 volts per Å is present between the two electrodes. Organic molecules subjected to this electric field can lose an electron by the tunnel effect to form positive ions. These are accelerated out of the source through the slit in the anode and then separated in the usual way by the mass spectrometer. Originally, f.i. sources suffered the serious drawback of very low sensitivity, but the activation of the emitter with benzonitrile has overcome this problem to some extent.[19]

Compared with the effort now being put into field desorption mass spectrometry, field ionization has received little attention of late, but the increased availability of field desorption sources which can be used readily for f.i. work may rectify matters. Although only a few references are available[20,21] to f.i.

mass spectrometry combined with gas chromatography, the method has great potential for quantitative work on complex mixtures, and may well compete with c.i. mass spectrometry in the near future, in view of the less frequent source cleaning required and superior reproducibility.

Field desorption

The use of benzonitrile as an activator leads to a large increase in the surface area of the emitter. For this reason, it becomes practicable to apply the compound directly to the emitter. The compound then undergoes field desorption from the emitter. The advantages of this method are that thermally unstable compounds tend not to decompose when using this technique, and high molecular weight compounds of low volatility can yield spectra. Not surprisingly, therefore, f.d. mass spectrometry has found application for the study of antibiotics,[22] polysaccharides[23] and especially in the sequencing of peptides.[24]

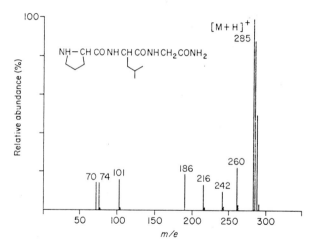

Fig. 1.6.   Mass spectrum of unmodified Pro-Leu-Gly-NH$_2$ by f.d. mass spectrometry. (Reproduced with permission from Ref. 24.)

Field desorption, being a soft ionization technique, is usually considered to yield only molecular weight information by virtue of the formation of [M]$^+$ and/or [M + H]$^+$ ions, depending upon the nature of the compound. However, frequently some structural information may be obtained, as, for example, in the spectrum of the tripeptide Pro-Leu-Gly-NH$_2$, shown in Fig. 1.6.[24] In this case it is possible to see fragment ions formed by cleavage at the amide bonds, thus giving extremely useful sequence information without the need for derivatization. It is to be expected that much more attention will be paid to the origin and possible structural significance of fragment ions in f.d. spectra in future.

Field desorption has been used recently as a quantitative technique. Early workers reported that even the isotopic abundances of chlorine containing compounds were not consistently in the ratio of 3:1, but these problems appeared to have been overcome. With a photoplate collector, it has been demonstrated that the ratio of diagnostic ions from a mixture of trisbansyldopamine and its [$^2H_4$] analogue are reproducible to within 5%.[25] It seems that the slow development of f.d. mass spectrometry as a quantitative technique is a reflection of the fact that so few ions are measured, leading to problems of ion statistics.

A report of the e.i., c.i. and f.d. spectra of some steroids[26] offers an opportunity to compare these three ionization techniques on the same compounds. Figure 1.7

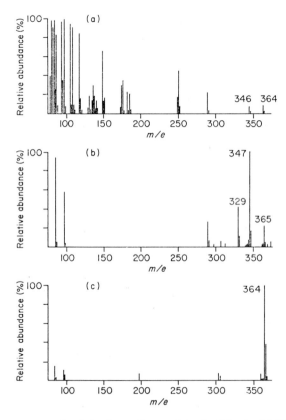

Fig. 1.7. Mass spectra of tetrahydrocortisone: (a) e.i.; (b) c.i.; (c) f.d. (Reproduced with permission from Ref. 26.)

shows the three spectra thus obtained for tetrahydrocortisone. In the e.i. mode the molecular ion is rather weak, and suffers the elimination of water. With methane c.i., the [MH]$^+$ ion is the base peak, but this shows considerable loss of water. At low wire currents [MH]$^+$ is the base peak in f.d. mode, but this changes to [M]$^+$ at high wire currents. Fragmentation is virtually absent.

Atmospheric pressure ionization

In this technique a carrier gas, usually nitrogen, is ionized at atmospheric pressure by radiation from a ^{63}Ni source. When a sample in solution, say in benzene, is injected, the benzene is ionized by charge exchange. The major reactant ions $[C_6H_6]^+$ and $[C_6H_7]^+$ ionize the compound by charge transfer and proton transfer, respectively, and the resulting ions have been analysed by a quadrupole mass spectrometer.[27] The instrument is extremely sensitive when in the single ion recording mode, detecting down to 150 fg of a dimethyl-γ-pyrone.[28]

Californium-252 plasma desorption

In this technique, developed recently by Macfarlane and Torgerson,[29] highly energetic fission fragments from the decay of ^{252}Cf are used to volatilize and ionize a solid sample. Each fission of ^{252}Cf produces two fragments travelling in opposite directions, and having unequal masses and energies. One fission fragment passes through a nickel foil upon which the solid sample has been deposited from solution, and, it is estimated, creates a local temperature of about 10 000 K, causing volatilization. Ionization occurs via ion–molecule reactions or ion pair formation. The ions are then passed down a time-of-flight tube. The fission fragment produced at the same time as the ionizing one, but which is travelling in the opposite direction, is detected by a scintillator and photo-multiplier and the signal used to provide a zero time to establish the mass scale for the ion travelling down the flight tube.

Since only a few ions are produced by each fission event of the ^{252}Cf, it is necessary to accumulate the results from a prolonged exposure to the fission fragments in order to produce a mass spectrum. This is conveniently done by a data acquisition system.

The method has great potential as a soft ionization process for involatile samples or samples which give no molecular ions under e.i. conditions, and tends to yield intense $[M + 1]^+$ ions. Compounds which have been studied by this technique include vitamin B_{12} and a number of neurotoxins not previously amenable to analysis by mass spectrometry using conventional e.i. sources.

ION SEPARATION

Single focusing magnetic instruments

A beam of ions of mass-to-charge ratio m/e is deflected on passage through a magnetic field according to the equation

$$m/e = H^2R^2/2V$$

where H is the strength of the magnetic field, R the radius of the circular path into which the ions are deflected, and V is the voltage used to accelerate the ions out of the source. A mass spectrum can be produced either by placing a photo-plate in the path of the ions (Fig. 1.8(a)) or by having a collector placed behind a narrow slit through which the ions must pass (Fig. 1.8(b)). In the first case all the ions are collected simultaneously. In the second case the magnetic field must be scanned downwards (or upwards) in order that successively decreasing (or increasing) m/e values achieve the correct path to pass through the collector slit, the accelerating voltage being kept constant. Alternatively, the magnetic field can be held constant while the accelerating voltage is scanned upwards or down-wards, allowing ions of successively increasing or decreasing m/e values to arrive at the slit. Magnetic scanning is used in most instruments since electric scanning can pose a number of problems. Thus, for example, the sensitivity of a mass spectrometer is related directly to the accelerating voltage, so that the sensitivity would decrease at high mass. Field penetration effects due to the accelerating voltage will also change as the latter is varied, causing some stability problems. These have been overcome for the most part in those instruments which are

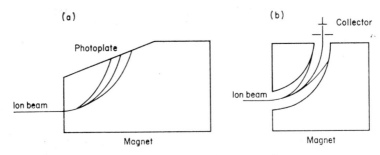

Fig. 1.8. (a) Collection of ions by a photoplate in a magnetic sector mass spectrometer. (b) Electrical detection of ions in a magnetic sector mass spectrometer.

commercially available. However, the technique of selected ion monitoring in magnetic instruments almost always depends upon switching the accelerating voltage between the values necessary to collect selected ions while the magnetic field is kept constant. This and other methods of selected ion monitoring are discussed later (p. 26).

A typical single focusing magnetic instrument scans from around m/e 10 up to about m/e 750 and has a resolving power of about 1000 (10% valley). Resolving power R is defined by

$$R = M/\Delta M$$

where M is the mean m/e value of two adjacent ions and ΔM is the difference between these ions. The allowable overlap between the two peaks is usually defined as 10% of the peak intensities, which are presumed to be equal. With a resolving power of 1000, doublets of m/e 100.00 and 100.10 would be separated to a 10% valley, while two ions of m/e 100.00 and 100.01 would be unresolved. Some manufacturers use a 50% valley between peaks for their statement of resolving power, and this should be borne in mind in any comparison of the resolving power of different instruments.

For combined gas chromatography mass spectrometry, a fast scanning instrument is essential. A typical g.c. peak obtained from a packed column may take 15 s to elute, and a 10-s scan would result in severe distortion of the spectrum compared with a 1-s scan. This is exemplified in Fig. 1.9, which shows the mass spectra of methyl stearate obtained through a g.c.m.s. system at two widely differing scan speeds. Some instruments have a capability for reducing the distortion by automatically changing the gain of the electron multiplier in accordance with the changing total ion monitor level, the latter being proportional to the sample level. This is not an ideal solution, however, since the normal constant background spectrum from source contaminants and column bleed

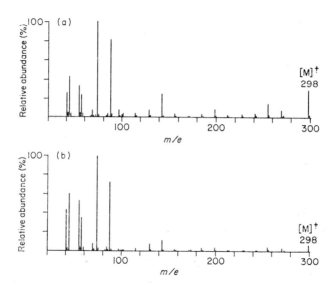

Fig. 1.9.   Mass spectrum of methyl stearate through a g.c.m.s. system: (a) at a scan speed of 3 s per decade; (b) at a scan speed of 20 s per decade.

will be distorted, resulting in problems when spectrum subtraction is attempted to clean up the background. Most modern instruments are capable of scan speeds of the order of 1–2 s per decade in mass.

A chart spectrum from a low resolution instrument yields m/e values to the nearest integer only. While a vast amount of information can be obtained from such a spectrum, especially with library matching facilities, the determination of the elemental composition of ions in the spectrum requires a measurement of their m/e values with an accuracy of a few parts per million. This is because different elemental combinations, although perhaps having the same integral mass, will differ slightly when their mass is measured to enough decimal places. To carry out such an accurate mass determination, it is necessary to have a reference compound, which is usually perfluorokerosene (PFK) or heptacosa-fluorobutylamine, bleeding into the source at the same time as the sample to be measured. Either a peak matching technique is used, which compares the two accelerating voltages necessary to bring the unknown ion and a convenient reference ion to a focus on the collector, or a complete scan can be run. A mass scale is established from the times of arrival of the reference peaks starting from the beginning of the scan, or the positions of the peaks on a photoplate if one is used. The times of the unknown peaks can then be converted to masses by inter-polation. Scanned accurate mass data is only really conveniently collected and processed by a computerized system that also subtracts the reference peaks from its printout and can calculate elemental compositions which are compatible with the determined m/e values.

The disadvantage of a low resolution mass spectrometer for carrying out accurate mass determinations is that it cannot resolve isobars, especially those involving sample ions and ions from the reference compound. Any measurements on such unresolved doublets or multiplets will be in error, and elemental com-positions will be meaningless. The m/e values of the highly fluorinated ions present in the spectrum are slightly less than the integral mass, while most organic compounds yield ions which are slightly higher in mass than the integral value. This means that a resolving power of about 10 000 is more than adequate to resolve PFK and sample peaks. This resolving power cannot be obtained with single focusing instruments, but requires a double focusing mass spectro-meter.

Double focusing mass spectrometers

The equation for ion separation given for single focusing instruments makes no provision for the fact that ions of the same m/e value can have a range of kinetic energies. In order to improve resolution, these ions should have a narrower band of kinetic energy. This state of affairs can be approached by passing the beam of ions through a radial electrostatic field, usually before the ion beam passes through the magnetic sector. The electrostatic field acts both as a direction and

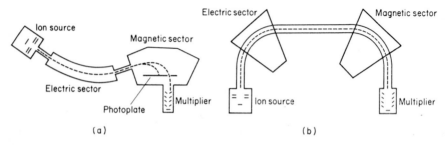

Fig. 1.10. (a) Schematic of a double focusing mass spectrometer with Mattauch–Herzog geometry. (b) Schematic of a double focusing mass spectrometer with Nier–Johnson geometry.

velocity focusing device. Directional focusing counteracts any angular divergence of the ion beam due to charge repulsion and the width of the ionizing electron beam. Two types of geometry are commonly used. In Mattauch–Herzog geometry (Fig. 1.10(a)) the ions are separated and focused onto a two-dimensional (planar) surface. Thus, a photoplate placed in the focal plane will record all the ions in the spectrum simultaneously. A point collector can also be used in electrical scanning. In Nier–Johnson geometry (Fig. 1.10(b)) the ions are separated and focused at a point. A collector is placed at this point. The mass spectrum is then obtained by scanning, as discussed for single focusing mass spectrometers.

Recent models of double focusing mass spectrometers are capable of static resolving powers, i.e. by peak matching, of 20 000 –100 000. In scanning mode, however, dynamic resolving powers are somewhat less than these, and are dependent upon scan speed and sample level, amongst other considerations. A resolving power of 10 000 at a scan speed of 10 s per decade should be easily attainable. As peaks become weaker the accuracy of mass measurement becomes less due to the collector receiving too few ions for a good peak shape. The peak shape deteriorates so that the calculation of the centroid becomes erratic (see Chapter 2). The accuracy of mass measurement in dynamic mode using a data acquisition system has been reported as being between 3.1 and 5.7 ppm for intense ions with the instrument set up carefully prior to determinations.[30]

The sensitivity of a mass spectrometer is inversely proportional to the resolving power used. Therefore, the instrument should not be operated at a higher resolving power than is necessary to resolve PFK from sample peaks, or resolve sample peaks which are themselves multiplets.

Double beam mass spectrometers

The A.E.I. MS-30 double beam double focusing mass spectrometer has two independent though adjacent sources. The ion beams from each source, although passing down the same flight tube, are kept separate, with only a few percent of cross-talk. Each ion beam has its own collector and electron multiplier. The shared flight tube means that both beams share the same electrostatic and magnetic fields. The separation of the two beams physically has several advantages, especially for g.c.m.s. work. There is no longer the problem of separating PFK ions from sample ions of the same nominal mass. Therefore, much lower resolving powers can be used, with consequent improvement in sensitivity. This is of considerable importance at the low sample levels encountered, say at the tens of nanograms level. The instrument can readily produce elemental compositions of all the ions in a mass spectrum obtained through a g.c. inlet with normal sample levels without overloading the g.c. column. Although no data are presently available, the double beam instrument should be of great value in determining elemental compositions with both f.d. and c.i. sources, since the reference source can still use PFK under e.i. conditions. Measurements have been carried out on f.d. spectra in single beam instruments by running PFK under e.i. conditions and then changing to sample under f.d. conditions.[31] A photoplate method has been described also using heptacosafluorobutylamine as reference,[32] which has yielded measurements accurate to ± 5 m.m.u.[33] In this case the volatile reference compound yields an f.i. spectrum, while the sample coated on the emitter gives an f.d. spectrum. A peak matching method has also been reported, with a mass measurement accuracy comparable to e.i. peak matching at 10 000 resolution.[34]

Quadrupole mass spectrometers

The quadrupole mass filter consists of four parallel rods arranged as shown in Fig. 1.11. Opposite rods are connected to radiofrequency (r.f.) and direct current (d.c.) voltages, one pair of rods being 180° out of phase with the other. Only ions of one particular m/e value can pass down the centre of the assembly for a particular value of r.f. and d.c. voltages. A mass spectrum is obtained by sweeping the voltages from a low value to a high value at a constant r.f./d.c.

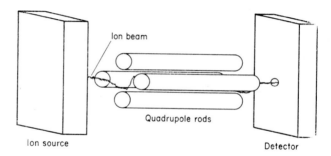

Fig. 1.11.   Arrangement of rods in a quadrupole mass spectrometer.

ratio. The mass scale is linear with respect to time. Quadrupoles have several advantages over magnetic instruments. They tend to be somewhat cheaper than magnetic instruments with similar mass ranges and resolution. In operation they are capable of rapid scanning, producing a spectrum (of poor quality) in about a millisecond if necessary, and, not requiring high voltage acceleration, can tolerate high source pressures, making them particularly suitable for g.c.m.s. work and chemical ionization. Switching of the rod voltages between various values necessary to focus a limited number of ions is obtained cheaply, so that selected ion recording is more readily automated.

ION DETECTION

Electron multipliers

Electron multipliers are used in the vast majority of mass spectrometers, even those of Mattauch–Herzog geometry equipped with photoplates. Multipliers have a fast response and high sensitivity, with gains up to 10^6. With such a gain, the signal-to-noise ratio for a single ion may be as much as 50:1.[35] The length of time for which a multiplier keeps a reasonable gain appears to be a matter of chance. One may lose a factor of 100 in a matter of six months while another might remain virtually unchanged. They can be nursed to a certain extent by keeping the multiplier voltage as low as possible. It is possible to clean some multipliers and restore part of their lost performance by washing in ethyl acetate followed by methanol. The device is then reactivated by heating in a g.c. oven overnight.

Daly collector

In the Daly collector[36,37] scintillation is produced by the positive ions impinging on an aluminium foil. The scintillation produces a response in a photomultiplier, the response being passed to the normal amplifying system of the mass spectrometer. The detector has the advantage that by the application of the correct voltages to various grids, normal ions can be suppressed and a response produced only for metastable ions.

Photographic recording

This is still used extensively in high resolution mass spectrometers with Mattauch–Herzog geometry. A typical photoplate will record about 30 mass spectra in rows. They have the advantage of being able to record a complete high resolution mass spectrum in seconds. Their capacity to integrate the spectrum is of great use in obtaining spectra from very small peaks emerging from a gas chromatograph. The high resolution spectrum which also contains the ions produced by the reference compound can be processed by densitometer and computer to produce measurements accurate to less than a millimass unit.[38] For quantitative work, a disadvantage is that emulsion response is not linearly proportional to the number of ions, so that a calibration curve of blackening versus sample level has to be produced for each plate.

SAMPLE ADMISSION

Direct inlet

Direct inlet probes are of either the indirectly or directly heated type. The sample, usually less than a microgram, is deposited in a quartz, glass, or gold cup. In the case of the indirectly heated probe, the probe is inserted through a vacuum lock into the source until it enters the ion chamber, which is typically at a temperature of 150–250 °C. The position of the probe is adjusted to give a suitable evaporation rate, as judged by the level of ions leaving the source. The latter is monitored by the total ion current recorder. The directly heated probe has a small heated block into which the sample tube fits, or makes connections inside the source. A stream of refrigerated air can be passed through the inside of the block in order to cool the probe and prevent flash evaporation. Once the instrument is set up the coolant can be turned off and the oven heated slowly as spectra are obtained.

Gas chromatograph effluents

The combination of a mass spectrometer with a gas chromatograph is complicated by the fact that the mass spectrometer source operates typically at about 10^{-7} mm pressure, whereas the gas chromatograph requires a pressure in excess

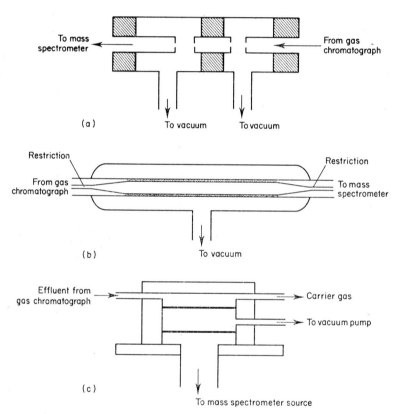

Fɪɢ. 1.12.   (a) Schematic of a jet separator. (b) Schematic of a glass frit separator. (c) Schematic of a membrane separator.

of 1 atm. In addition, the sample is diluted by the carrier gas by a factor of perhaps many thousands. Various types of molecular separators have been developed to overcome these problems. Their function is to remove the carrier gas in preference to the sample. Important considerations with separators are the yield, Y, the enrichment, N, and the effect on the resolution and shape of the g.c. peaks. The enrichment can be defined as the ratio of sample concentration in the carrier gas entering the mass spectrometer to the sample concentration in the effluent at the end of the g.c. column. The amount of sample actually entering the source is a more important consideration than the enrichment, so that the yield Y, defined as follows, gives a more meaningful comparison of separators:

$$Y = N \frac{Q_{\text{m.s.}}}{Q_{\text{g.c.}}}$$

where $Q_{\text{m.s.}}$ is the carrier gas flow into the mass spectrometer and $Q_{\text{g.c.}}$ is the carrier gas flow into the column.

Of the separators discussed in the following paragraphs, only the membrane type is liable to cause severe peak broadening and loss of resolution, and this usually occurs because of a marked difference in temperature between the effluent from the gas chromatograph and the separator itself.

The Ryhage separator[39] shown in Fig. 1.12(a) is made of stainless steel and consists of two separately pumped compartments, each with two opposed jets through which the effluent passes on its way to the mass spectrometer source. The first compartment, pumped by a rotary pump operates at about 0.1 Torr, while the second is pumped by a diffusion pump and is at a pressure of about 10^{-3} Torr. These pressures are with flow rates of about 30 ml min^{-1}. The efficiency of a two-stage separator (the ratio of sample leaving to sample entering the separator) at about 60% is higher than that of a single-stage version. The latest single-stage separators have the great advantage of being made in glass, however, thus eliminating problems due to thermal decomposition being catalysed by the stainless steel.

With a single-stage jet separator, it is probable that the rapid pressure drop means that the last portion of the g.c. column is operating under vacuum. This might cause some loss of resolution of the g.c. peaks or change retention times, but does not appear to be a particularly serious problem.

The Watson–Biemann separator[40] (Fig. 1.12(b)) is used quite widely. This consists of a porous glass tube mounted in a glass envelope which is pumped by a rotary pump. The g.c. effluent passes through the porous tube and into the mass spectrometer source. The porous tube has a restriction at either end to achieve the necessary pressure reduction for molecular flow to occur. The large surface area of the glass in the separator would appear at first to be a problem, since polar compounds would be adsorbed on the surface. However, this can be avoided by silanizing the separator with, for example, bis(trimethylsilyl)acetamide. This is best carried out by removing the separator, since the silanizing agent may cause harmful deposits in the mass spectrometer source. A typical separator of this type has an efficiency of about 30%.

The permeable membrane separator[41] (Fig. 1.12(c)) consists of a silicone rubber membrane supported on a glass frit, the membrane being the interface between the spectrometer and the gas chromatograph. The g.c. effluent is passed over the surface of the membrane, which is kept within 30 °C of the column temperature. The upper temperature limit of the membrane is about 230 °C.

The organic compound dissolves in the membrane and passes through to the mass spectrometer source. The amount of carrier gas passed by the separator is of the order of 0.2 ml min^{-1}, and source pressures of about 5×10^{-6} Torr can be obtained with a single-stage separator. The efficiency is of the order of 50%, and increases with increased carrier gas flow. Membrane separators cause considerable tailing of high boiling polar compounds, and this can sometimes become unacceptable.

The above separators all give an optimum yield for one particular flow rate of carrier gas. However, a variable slit separator devised by Brunnée overcomes the problem.[42] The variable effusion area is a circular slot formed by two concentric knife edges with a glass plate lying on top. The distance between the knife edges and the plate can be varied to give optimum performance for the particular flow rate used for the g.c. column.

All separators work best with a carrier gas of the minimum molecular weight. This also has the advantage that ions due to the carrier gas occur at a part of the spectrum that is of little importance to organic chemists. Helium is now used virtually exclusively as carrier gas in g.c.m.s. systems. A further advantage with helium is that it has a high ionization potential (about 24 eV). If the beam energy of the mass spectrometer is kept at 20 eV during recording of the total ion current, the helium will make virtually no contribution, so that greater sensitivity to small amounts of organic compounds eluting into the source will be obtained on the total ion current chromatogram. If it is desired to scan a spectrum at that point, the electron voltage can be switched rapidly to the normal operating one of 70 eV.

A full description of most of the separators in use up to 1973 can be found in the comprehensive book on g.c.m.s. by McFadden.[43]

Column bleed

In gas chromatography the problem of high bleed from columns can be overcome for example by the use of dual column instruments. One column is used for the analysis while the signal produced by bleed can be offset by the same signal produced by the second column operating under the same conditions. Unfortunately this system cannot be used in single source g.c.m.s. systems. With a computerized data acquisition system, the problem can be surmounted by the use of spectrum subtraction or the maximizing mass technique discussed in Chapter 2.

In the case of simple spectrum subtraction, the improvement which can be obtained is shown in Fig. 1.13.

If a computer is not available, the spectrum subtraction technique is too tedious to be carried out manually, since the raw spectrum and background spectrum intensities must be adjusted to give the same intensity for a designated background peak prior to subtraction. A practical solution may be to use a short length of bleed absorbing column at the end of the main column.[44] The bleed absorbing column contains a more stable stationary phase than the main column,

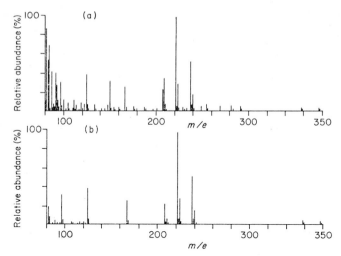

Fig. 1.13. Mass spectrum of haloperidol through a g.c.m.s. system: (a) raw spectrum; (b) after subtraction of the background.

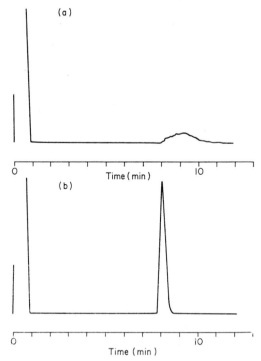

Fig. 1.14. Total ion monitor trace following the injection of 100 ng of mepyramine in a g.c.m.s. system: (a) with a number of cold spots between the column and the mass spectrometer source; (b) with an even temperature between the column and mass spectrometer source.

but, due to its short length, has a negligible effect on the resolution of components emerging from the main column. Sometimes the system can result in a quite remarkable reduction in the bleed level entering the mass spectrometer.

Low bleed columns are becoming much more widely available, so that it is fairly easy to find an alternative column to reduce the problem of bleed.

Cold spots

It is of the utmost importance that the temperature of the column, the separator and the inlet line to the mass spectrometer source should be either constant or exhibit only a slight gradient from one end to another. Any cold spot present will act to the detriment of the system by causing condensation of the more involatile components. Severe cold spots may even prevent a component reaching the mass spectrometer source unless particularly high column loading is used. The latter may lead to degradation of column resolution. Both of these effects are illustrated in Fig. 1.14, in which the total ion current trace is given for an injection of 100 ng of mepyramine, first for a stainless steel jet separator with stainless steel tubing to the mass spectrometer source, the system containing several swagelock connections, and second, where these had been removed.

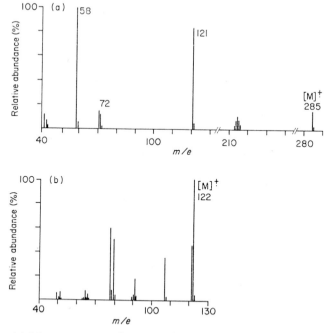

Fig. 1.15.   (a) Mass spectrum of mepyramine scanned at the top of the peak shown in Fig. 1.14(b). Stainless steel inlet line to the mass spectrometer source at 200 °C. (b) Mass spectrum obtained with the inlet line at 210 °C. The spectrum is identical to that of an authentic sample of *p*-methoxytoluene.

A remarkable improvement is obtained by the removal of connectors, which are notorious for producing cold spots.

Stainless steel or other metals are gradually being phased out of g.c.m.s. systems because of the problems they cause with thermally sensitive compounds. The first spectrum in Fig. 1.15 was obtained from an injection of mepyramine with the line between separator and mass spectrometer being kept at 200 °C. When the temperature of the line was raised by only 10 °C, the spectrum in Fig. 1.15(b) was obtained, corresponding to *p*-methoxytoluene. Since the retention time of this compound on the system was identical to that of mepyramine on the same system, it was obvious that decomposition was taking place in that portion of the interface between the separator and the source. When a monohydroxylated derivative of mepyramine was injected into the same system, no response was obtained at a range of inlet temperatures. However, when a glass jet separator with all glass inlet lines was employed, reasonable mass spectra were obtained for this and other multihydroxylated derivatives.

REFERENCES

1. F. W. McLafferty, T. Wachs, C. Lifshitz, G. Innorta and P. Irving, *J. Am. Chem. Soc.* **92**, 6867 (1970).
2. G. P. Arsenault, in *Biochemical Applications of Mass Spectrometry* (G. R. Waller, editor), Wiley, New York, 1972, p. 819.
3. D. M. Desiderio and K. Hagele, *Chem. Commun.* 1074 (1971).
4. G. W. A. Milne, H. M. Fales and T. Axenrod, *Anal. Chem.* **43**, 1815 (1971).
5. D. F. Hunt and J. F. Ryan, *Anal. Chem.* **44**, 1306 (1972).
6. J. Yinon, *Biomed. Mass Spectrom.* **1**, 393 (1974).
7. A. M. Hogg and T. L. Nagabhushan, *Tetrahedron Lett.* 4827 (1972).
8. G. P. Arsenault, J. J. Dolhum and K. Biemann, *Chem. Commun.* 1542 (1970).
9. R. L. Foltz, *Lloydia* **35**, 344 (1972).
10. M. G. Horning, J. Nowlin, K. Lertratanangkoon, R. N. Stillwell, W. G. Stillwell and R. M. Hill, *Clin. Chem.* **19**, 845 (1973).
11. F. H. Field, *Acc. Chem. Res.* **1**, 42 (1968).
12. F. H. Field, in *Advances in Mass Spectrometry*, Vol. 4 (E. Kendrick, editor), The Institute of Petroleum, London, 1968, p. 645.
13. F. H. Field, in *Mass Spectrometry* (A. Maccoll, editor), MTP International Review of Science, Series 1, Vol. 5, Butterworths, London, 1972.
14. B. L. Jelus, B. Munson and C. Fenselau, *Biomed. Mass Spectrom.* **1**, 96 (1974).
15. B. L. Jelus, B. Munson and C. Fenselau, *Anal. Chem.* **46**, 729 (1974).
16. I. Jardine and C. Fenselau, *Anal. Chem.* **47**, 1730 (1975).
17. C. Fenselau, *Appl. Spectrosc.* **28**, 307 (1974).
18. H. D. Beckey, *Field Ionisation Mass Spectrometry*, Pergamon Press, Oxford, 1971.
19. H. D. Beckey, E. Hilt, A. Maas, M. D. Migahad and E. J. Ochterbeck, *Int. J. Mass Spectrom. Ion Phys.* **3**, 169 (1969).
20. J. N. Damico and R. P. Barron, *Anal. Chem.* **43**, 17 (1971).
21. D. E. Games, A. H. Jackson, D. S. Millington and M. Rossiter, *Biomed. Mass Spectrom.* **1**, 5 (1974).
22. K. L. Rinehart Jr, J. C. Cook Jr, K. H. Maurer and U. J. Rapp, *J. Antibiot.* **27**, 1 (1974).

23. W. D. Lehmann, H.-R. Schulten and H. D. Beckey, *Org. Mass Spectrom.* **7**, 1103 (1973).
24. S. Asante-Poku, G. W. Wood and D. E. Schmidt Jr, *Biomed. Mass Spectrom.* **2**, 121 (1975).
25. W. D. Lehmann, H. D. Beckey and H.-R. Schulten, *Anal. Chem.* **48**, 1572 (1976).
26. N. J. Haskins, D. E. Games and K. T. Taylor, *Biomed. Mass Spectrom.* **1**, 423 (1974).
27. E. C. Horning, M. G. Horning, D. I. Carrol, I. Dzidic and R. N. Stillwell, *Anal. Chem.* **45**, 936 (1973).
28. D. I. Carrol, I. Dzidic, R. N. Stillwell, M. G. Horning and E. C. Horning, *Anal. Chem.* **46**, 707 (1974).
29. R. D. Macfarlane and D. F. Torgerson, *Science*, **191**, 920 (1976).
30. R. S. Gohlke, G. P. Happ, D. P. Maier and D. W. Stewart, *Anal. Chem.* **44**, 1484 (1972).
31. H.-R. Schulten and D. E. Games, *Biomed. Mass Spectrom.* **1**, 120 (1974).
32. H.-R. Schulten and H. D. Beckey, *Org. Mass Spectrom.* **7**, 861 (1973).
33. H.-R. Schulten and U. Schurath, *J. Phys. Chem.* **79**, 51 (1975).
34. C. N. McEwen and A. G. Bolinski, *Biomed. Mass Spectrom.* **2**, 112 (1975).
35. B. N. Green, T. O. Merren and J. G. Murray, American Society for Mass Spectrometry, 13th Annual Conference on Mass Spectrometry and Allied Topics, St Louis, Missouri (1965). Proceedings, p. 204.
36. N. R. Daly, A. McCormick and R. E. Powell, *Rev. Sci. Instrum.* **39**, 1163 (1968).
37. N. R. Daly, A. McCormick and R. E. Powell, *Org. Mass Spectrom.* **1**, 167 (1968).
38. J. T. Watson and K. Biemann, *Anal. Chem.* **37**, 844 (1965).
39. R. Ryhage, *Anal. Chem.* **36**, 759 (1964).
40. J. T. Watson and K. Biemann, *Anal. Chem.* **36**, 1135 (1964).
41. P. M. Llewellyn and D. Littlejohn, Conference on Analytical and Applied Spectroscopy, Pittsburgh (1966).
42. C. Brunnée, L. Delgmann, K. Habfast and S. Meier, American Society for Mass Spectrometry, 18th Annual Conference on Mass Spectrometry and Allied Topics, San Francisco, California (1970). Proceedings, p. B306.
43. W. McFadden, *Techniques of Combined Gas Chromatography/Mass Spectrometry*, Wiley, New York, 1973.
44. R. L. Levy, H. Gesser, T. S. Herman and F. W. Hongen, *Anal. Chem.* **41**, 1480 (1969).

CHAPTER 2

FURTHER ASPECTS OF INSTRUMENTATION

METASTABLE TRANSITIONS

It has already been stated that ions spend about 10^{-6} s in the ion source. On leaving the source after acceleration, ions typically take about 10^{-5} s to traverse the length of the flight tube to the collector. When decomposition occurs with a rate constant of less than 10^5 s^{-1}, the parent ion reaches the collector before decomposition. With rate constants of greater than 10^6 s^{-1}, decomposition takes place before acceleration out of the source, and the daughter ion is collected. Metastable transitions are decompositions which occur between the source and collector, i.e. those with rate constants between 10^5 and 10^6 s^{-1}. For the decomposition $m_1^+ \rightarrow m_2^+$ occurring in the field-free region before the magnet in the case of a single focusing instrument and between the electric and magnetic sectors in a double focusing spectrometer, a broad peak is observed centred at the apparent m/e value $m^* = m_2^2/m_1$. Thus, of course, the presence of a metastable ion confirms the formation of m_2^+ from m_1^+. In double focusing instruments, another field-free region of particular interest for the observation of metastable ions occurs between the source and the electrostatic field. If the transition $m_1^+ \rightarrow m_2^+$ occurs in this region, the m_2^+ ions have a translational energy of $(m_2/m_1)V_0e$, where V_0 is the accelerating voltage and e the electronic charge. During normal operation, the ratio of the accelerating voltage to the electrostatic sector voltage E_0 is kept constant, e.g. about 14.8 for the A.E.I. MS-902. At this ratio, ions with translational energies in the range $V_0e(1 \pm 0.0066)$ are allowed through the electrostatic sector. The m_2^+ ion with energy $(m_2/m_1)V_0e$ will obviously be outside this range. However, if the ratio V_0/E_0 is changed to a value $V/E = (m_1/m_2)V_0/E_0$, the ion m_2^+ will then be transmitted, while normal ions will not be allowed through. This is the method of metastable refocusing mentioned earlier,[1,2] and since the ratio V/E can be measured accurately, the precursors of the ion m_2^+ can be determined if the spectrometer is tuned to collect $m_2{}^+$ prior to changing V/E. If a computer is used, a fragmentation map can be determined for a particular compound.[3] This can be of great importance in structural determination. In ion kinetic energy spectroscopy (i.k.e.s.) an electron multiplier is placed at the exit slit from the electrostatic analyser so that a spectrum of ion kinetic energies for ions entering the analyser can be determined. A useful application of this technique has been to distinguish between two isomers which may yield virtually identical normal mass spectra.[4] However, it should be noted that the technique does not always work.[5]

In reversed geometry double focusing instruments, i.e. those with the magnetic analyser preceding the electrostatic analyser, a particular ion can be selected by the magnetic analyser for study by sweeping the electrostatic analyser voltage. Both of these ion kinetic energy spectroscopic techniques may have a possible application in quantitative work for the determination of mixtures of isomeric compounds which individually give similar mass spectra.

SELECTED ION MONITORING

Various terms have been used for the process in which only a few m/e values, say one to eight, are observed continuously as a function of time instead of scanning the complete spectrum.[6] Mass fragmentography, the term first used by Hammar et al.,[7] is rather unsatisfactory, because in many instances molecular ions rather than fragment ions are observed. The term is even less applicable to c.i. mass spectrometry.

In a conventional scan of, say, 2 s per decade, each mass in the spectrum is focused on the collector for only a few milliseconds, the exact time of course depending on the resolution at which the instrument is operating. A satisfactory response can be obtained from ions of low intensity only when samples of the order of 10 ng to 1 µg are used. If only one m/e value is monitored continuously, it is focused at the collector for the whole period of elution of a g.c. peak where a g.c.m.s. system is considered. Since a g.c. peak on a packed column is typically of the order of 15 s wide, the number of individual ions of the particular m/e value monitored which are received by the collector will be several thousand times the number that would be received in the 2-s scan. Therefore, the technique of single ion monitoring can be thousands of times more sensitive for a particular compound than would be the measurement of that ion from a scanned spectrum. If, for example, four ions were monitored continuously, the technique would be expected to be perhaps a thousand times more sensitive than scanning mass spectrometry.

Any mass spectrometer can be set up to record just one ion continuously, and this technique was used as long ago as 1959 by Henneberg.[8,9] To monitor more than one mass requires the use of special equipment, which tends to be costly as far as magnetic instruments are concerned. Selected ion monitoring is a comparatively straightforward matter in the case of quadrupole instruments, since the voltages on the quadrupole rods are relatively low. There is no limitation except sensitivity on the masses which can be monitored during one run. For example, ions as far apart as m/e 50 and m/e 500 can be selected. The voltages on the rods are switched sequentially between those necessary to focus the chosen ions, and the signals are output via a sample and hold amplifier circuit to a multipen recorder in the absence of a data system. As many as eight ions have been recorded simultaneously.[10] Computer-controlled selected ion monitoring in quadrupole g.c.m.s. systems is now becoming commonplace.

From the equation $m/e = H^2R^2/2V$ given in Chapter 1 for the separation of ions by magnetic instruments, it can be seen that selected ion monitoring is possible either by changing the magnetic field rapidly between the values necessary to focus the required ions or by changing the accelerating voltage rapidly. However, magnetic fields produced by large electromagnets such as are used in mass spectrometers take a comparatively long time to rise or decay, so it is not practicable to use magnetic switching for selected ion recording. In the first report of selected ion monitoring in a magnetic instrument coupled to a gas chromatograph, Sweeley et al.[11] used a device called an accelerating voltage alternator (a.v.a.). This switched the accelerating voltage between the two values necessary to focus two ions simultaneously, and was used to resolve a mixture of glucose and [^{2}H$_7$]glucose which eluted from the gas chromatograph with the same retention time. Subsequently the device was improved to record three ions simultaneously.[7]

More important than the mass range over which the a.v.a. can function is the question of stability. The voltages are adjusted so that the tops of the peaks are focused at the collector. However, drift may occur from the tops of the peaks due to instability of the instrument electronics. This problem worsens as the resolving power of the instrument is increased and the peaks become sharper, so that the tendency is to work at low resolving powers with wide slits whereby peaks become somewhat flat-topped. It is possible to use a computer to correct for this drift by adjusting the accelerating voltage continually through the addition or subtraction of a small offset voltage.[12] Klein[13,14] has used the concept of adjusting the accelerating voltage to a value just short of that necessary to focus the ion, and then adding a small offset voltage generated by a linear sweep generator synchronized with the switching cycle of the a.v.a. By this means the complete m/e peak is swept out, and can be displayed on an oscilloscope. It is a great advantage to be able to visualize the peak shape in this way during a run, since any drift can be compensated for immediately to bring the peak into the centre of the screen. Another advantage is that the presence of a multiplet can be seen easily, and action taken to improve the resolving power if this is possible to completely separate the components of the multiplet and refocus upon the ion with the correct elemental composition.

Gruenke and Craig[15,16] use the technique of superimposing an alternating current (a.c.) voltage upon the voltage chosen by a two-channel a.v.a. Besides the advantage of its low cost, the variable voltage a.c. signal (typically 60 Hz) allows the chosen peak to remain centred on the oscilloscope while the mass range swept is narrowed or widened. By narrowing the sweep on multiplets, it can be arranged that only the ion of the desired elemental composition is recorded, thus avoiding contributions from the background ions at the same nominal mass.

COMPUTERIZED DATA ACQUISITION

Computers first came into mass spectrometry about twelve years ago for the calculation of elemental compositions of all the ions in a high resolution spectrum recorded on a photoplate.[17] Shortly afterwards electrically scanned spectra were recorded on magnetic tape, played back through an analogue-to-digital converter (a.d.c.), and processed by a computer to give accurate masses and hence elemental compositions of all the ions in the mass spectrum.[18] Accuracies of the order of 10 ppm were obtained. Both of these systems were *off-line*, i.e. at some stage there was storage of data before processing by the computer. Probably the first on-line system in which the mass spectrometer was connected to the computer via an a.d.c. was that reported two years later.[19] Thus, the scene was set for the revolution that has occurred since in the development of computer systems dedicated to a mass spectrometer. Until about 1970, computer systems tended to be bought in order to realize the maximum potential of high resolution mass spectrometers. The peak matching 'manual' method of accurate mass determination, even in highly skilled hands, is capable only of measuring perhaps one peak a minute in a spectrum, although this is the most accurate method of measurement. To process a complete spectrum in this way is obviously tedious, yet a computer can produce elemental compositions for a complete scan in minutes if a line printer or visual display unit is available as the output device.

The vastly increased use of g.c.m.s. both in a qualitative and quantitative sense has resulted in recognition of the fact that computers are also essential for low resolution mass spectrometry. The techniques used in low resolution data acquisition systems are somewhat different from those used in high resolution work.

HIGH RESOLUTION DATA ACQUISITION

A block diagram of a typical high resolution data acquisition system is shown in Fig. 2.1.

The multiplexer is a device that enables several input signals to be sampled by the a.d.c. Thus it can be used for taking the output from several mass spectrometers to the same computer. More obvious is its use as a device to sample both the spectrum and the total ion current monitor of the mass spectrometer at the same time. In a g.c.m.s. system it is possible also to sample the output from a separate gas chromatographic detector where this is used.

Since digital computers obviously require a digital input, an a.d.c. is necessary to change the analogue signal from the mass spectrometer amplifying system into

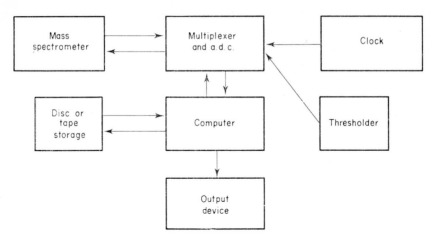

Fig. 2.1.   Block diagram of a typical data acquisition system for a high resolution mass spectrometer.

digital information. Two parameters of the a.d.c. are important, the number of bits into which the analogue signal is converted, and the digitization rate. Bit capacities of from 10 to 14 and digitization rates of from 50 to 250 kHz are fairly typical. The effect of these parameters on error is discussed in Chapter 3.

The thresholder is necessary to avoid the situation of the computer processing the areas of no information that occur between mass spectral peaks during a scan. In a typical scan peaks are observed for only a few percent of the scan period. The threshold device allows information to be passed to the computer only if the signal level rises above a preset value, i.e. when a mass spectral peak occurs. Noise spikes can rise above the threshold, and the computer is usually programmed to deal with these by having a proviso that three or four consecutive samples must rise above the threshold before they are processed.

Until recently, the majority of small dedicated computers were of the 12-bit 8K core type, since the job they were required to do was relatively simple. Larger computers of 16 and 32K are now used more frequently, first because 8K is normally insufficient to drive a visual display unit, and second because there is a great degree of sophistication in the tasks which the computer is required to carry out, for example library searching. A fast 32K 16-bit computer should prove adequate to cope with recent and future developments in the manipulation of data.

Where data output is concerned, the teletype is considered to be much too slow, since it can take as long as half an hour to output accurate mass and intensity data for a medium molecular weight compound. Line printers, visual display units, printer-plotters and incremental plotters are now preferred, while, as storage devices, disks capable of holding more than a million words have superseded magnetic tape.

The usual method of obtaining high resolution data is to obtain scans at a dynamic resolving power of, say, 10 000 on a mixture of PFK and the compound. The voltage output from the mass spectrometer is sampled at a constant high rate, such as every 50 μs. As mentioned previously, when the voltage exceeds a given threshold for a number of consecutive samples, a peak is defined. The threshold can be set either by the thresholder or by means of software. The computer has then to calculate the centre of gravity (centroid) of the peak. As shown for the typical peak in Fig. 2.2, what the computer receives is a series of

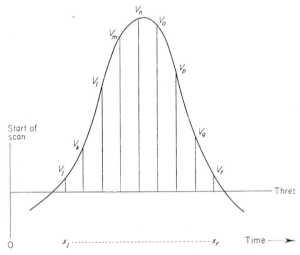

Fig. 2.2.   Digitization of a mass spectral peak. The voltages V_j to V_r are sampled at times x_j to x_r from the start of the scan.

times x_j to x_r from the beginning of the scan, and the corresponding voltages V_j to V_r at these sampling points. The centroid C is given by

$$C = \frac{\sum_{n=j}^{r} V_n x_n}{\sum_{n=j}^{r} V_n}$$

It is necessary also to have an intensity measurement for each peak, and this is given by the expression

$$\sum_{n=j}^{r} V_n$$

After this preliminary calculation a list is obtained of times and intensities of all the peaks in the spectrum of PFK plus compound. These are compared with a list of times for a scan of PFK by itself so that eventually the spectrum of the pure compound can be obtained. In order to convert these times to accurate masses it is necessary to tell the computer the times of the first three PFK peaks. Using a modified exponential relationship, the computer searches for the next PFK peak and carries out a correction to take into account the difference between the observed and calculated time. This process continues down the spectrum until all the reference peaks have been accounted for, although most programmes allow several reference peaks to be missed. The times of the compound peaks can be converted into masses by interpolation using the relationship calculated for the PFK peaks.

In another method developed by Carrick[20,21] it is argued that it is not necessary to sample a peak at short time intervals in order to determine its profile and hence its centroid. According to Carrick, peaks are triangular in shape, and it is necessary only to determine the peak maximum. This is carried out by a derivatizing circuit. The device can be obtained in an off-line form which produces a paper tape suitable for computer processing, or on-line with a dedicated computer. Its performance seems to be comparable with a centroid type of interface, both of these methods being capable of better than ± 10 ppm accuracy at a scan rate of 10 s per decade.

The use of a densitometer to obtain data suitable for computer processing from high resolution photoplate spectra was pioneered by Biemann.[22] More recently the same Mattauch–Herzog instrument was used to compare the accuracy of mass measurement of both the photoplate and scanning methods.[23] It was shown that the photoplate method gave more accurate results and needed 20 times less sample. The greatest drawback of the photoplate method is its inconvenience and timelag in processing data. For a large g.c. run it is impossible to change the photoplate in time to record all the data.

LOW RESOLUTION DATA ACQUISITION

If only integral mass values are required, it is not necessary to have the PFK reference compound present at the same time as the sample. At the beginning of the day several scans are taken with only PFK in the spectrometer. These reference scans provide a mass calibration for future scans started from the same point, so that it is possible to subsequently dispense with the PFK. The calibration frequently holds good for several days with a stable mass spectrometer.

In the case of the double beam instrument, PFK is bled continually into one source, and even at low resolution, a calibration is held to within about 10 ppm for the rest of the day.

COMPUTER CONTROL OF MASS SPECTROMETERS

The concept of computer control of the mass spectrometer was first applied to quadrupole instruments, since these presented fewer problems than magnetic instruments. Perhaps the first complete system was that described by Reynolds *et al.*[24] The computer was programmed to generate the voltage applied to the quadrupole rods, the voltages being obtained from a calibration run. Only the voltages necessary to focus each mass were applied sequentially, so that no time was wasted in scanning portions of the spectrum in which no information resided. The two advantages of this approach are that no computer time is wasted in processing useless data, and more time is spent sampling each mass spectral peak, with consequent improvement in sensitivity and signal-to-noise ratio.

The latest commercial instruments employ a similar system.[25] The mass spectra consist of counts per channel, a channel being one a.m.u. Various functions can be performed by push-button control, such as spectrum subtraction and integration. With this system it is possible to progress from a full scan down to single ion monitoring.

When using selected ion recording, most workers think in terms of four or six ions, since this number usually removes any ambiguity as to the identity of the compound being monitored. However, Green and Hertel[26] use the concept of a contracted mass spectrum, in which 10 to 15 ions are monitored under computer control. The number of ions monitored depended upon the uniqueness of the spectrum of the compound concerned. With this large number of ions, it is claimed that the extra selectivity given by the degree of retention of a compound by a g.c. column is not required, so that a flash evaporator with a membrane separator could be used as the inlet system. By this means a rapid analysis, taking perhaps a few seconds, could be carried out on 'street' drugs and urine extracts.

In the case of magnetic instruments, the use of computers appears to have been restricted mainly to selected ion recording rather than complete scanning of the spectrum. The problem of stability in accelerating voltage alternators has been mentioned earlier, and one application of computers has been to continually add or subtract a small offset voltage to the accelerating voltage.[14] By this means it can be ensured that the maximum signal is received by the collector. In an application to the measurement of prostaglandins using deuterium-labelled

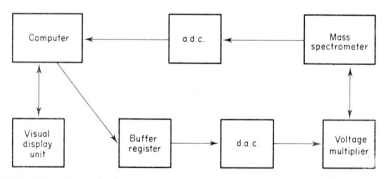

Fig. 2.3. The main features of a computer controlled mass spectrometer.

internal standards it was possible to determine isotopic abundances with a precision of greater than 1 %.

For greater computer control, it is necessary to regulate several voltages. The accelerating voltage must be set, and it is necessary also to be able to offset the signals due to background at any particular mass. The selected ion recording should be displayed in real time. A typical block diagram of the main features of such a system is shown in Fig. 2.3.

In a system described by Elkin et al.,[27] a 14-bit digital-to-analogue converter (d.a.c.) was used to set the accelerating voltage of the mass spectrometer for the required masses by adding voltages from 0 to 1000 volts to the existing accelerating voltage. This gave a mass range of 30 % between the lowest and highest selected voltages. An 8-bit d.a.c. was used to provide voltages to offset any background at a particular mass. The various operating parameters were set up on a visual display unit. Another laboratory has used a system which can monitor up to twelve different masses.[28,29] These covered a mass range of 40 %, the accelerating voltage being controlled stepwise between 2·5 and 3·5 kV.

An alternative to selected ion recording is to use the computer to scan the magnet repetitively through a small mass range.[30] Using a range of about 20 mass units, levels of drugs down to the hundreds of picograms were determined. This appears to be about a hundred times less sensitive than true selected ion recording (see Chapter 6), and is expected to be so because less time is spent collecting ions at each mass position. However, it has been argued that with correct setting up of the system, repetitive scanning over a limited mass range is of similar sensitivity to selected ion recording, although recent improvements in the technique of selected ion recording make this seem rather unlikely (Chapter 6).

REPETITIVE SCANNING OF FULL SPECTRA

The use of a mass spectrometer in repetitive scanning mode at a fast rate such as 3 s per decade has become possible only with the advent of data acquisition systems with fairly large storage devices. The technique was reported originally by Hites and Biemann in 1970,[31] who used the term 'mass chromatography' to indicate that the elution profile of any ion in the g.c.m.s. combination could be determined. From the point of view of flexibility, the repetitive scanning mode is selected ion recording *par excellence*, since it is necessary only to carry out one sample injection and data acquisition run. Selected ions can be retrieved from the scan data stored on disk. In a conventional selected ion recording experiment, it is frequently necessary to carry out numerous injections of sample monitoring different combinations of ions each time until the best conditions for the analysis are found. Under adverse conditions this may take a whole day. The repetitive scanning method allows a rapid evaluation of the most appropriate ions for the analysis, and an example is given in Fig. 2.4.

As well as the ability to select any ion in the spectrum, each ion profile can be displayed with a different gain, allowing even greater flexibility. Most com-

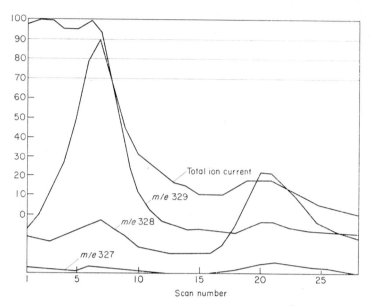

Fig. 2.4.   Retrieval of the ions *m/e* 327, 328 and 329, and total ion current from
repetitive scanning with a data acquisition system.

mercial data acquisition systems now offer selective ion retrieval from repetitive
scan data. All the advantages outlined above must be paid for, however, by the
major disadvantage of loss in sensitivity compared with selected ion recording.
The technique appears to be a thousand times less sensitive than the continuous
monitoring of a single ion under static mass spectrometric conditions.

Repetitive scanning has been used by Biller and Biemann[32,33] to produce
'mass resolved chromatograms'. The method depends upon finding all those ions
that maximize in intensity at a particular scan. A peak profile analysis is performed for each mass chromatogram to identify all those scans where that mass
maximizes by using an algorithm analogous to that employed to find the peak
centroids during normal data acquisition. This results in the formation of an
array for each scan, containing those *m/e* values that maximize, together with
their abundance. These resulting spectra retain the characteristic peaks of the
spectrum but have eliminated the contributions from incompletely resolved
components. A computer match with a library of spectra has a much better
chance of identifying the component than if the original data from the scan
were used.

A further advantage of the method is that the sum of the intensities of only
the ions that maximize in a given scan produces a much better resolved total ion
current chromatogram than the original data. This can be seen in Fig. 2.5,
which compares a total ion current chromatogram with the mass resolved
chromatogram.

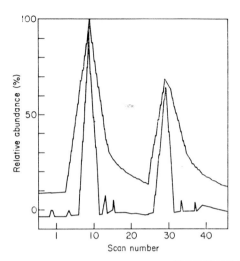

Fig. 2.5. Comparison of total ion current chromatogram (upper trace) with mass resolved chromatogram (lower trace).

The improvement in resolution increases as decreasing scan rates are used, so that the best results are obtained at scans of 1 s per decade. With a rapidly changing total ion current and slower scans, there is a danger that high and low mass ions in the same scan may be judged to be maximizing in adjacent scans, so becoming resolved when they should not. Another possibility is that masses common to two components of a partially resolved g.c. peak may be assigned to only one of these components.

Duffield et al.[34] have used an approach which starts by computing two histograms. The first measures the number of single ion profiles that reach maximum in each time interval of the scan, while the second measures the ion intensity above background in these maxima. At a given elution time the histograms include peak maxima from all masses over a spread of seven spectra. The position of each ion maximum is determined by a parabolic least squares interpolation about the top five points in the peak data. The time coordinates of the maxima are then estimated to one-third of a scan time. By this method eluants coming off the column within $1\frac{1}{2}$ to 2 spectral scan periods of each other can be distinguished readily.

LIBRARY SEARCHING

The facility to search a library of spectra by computer for a match with an unknown spectrum is now routinely offered by all manufacturers of mass spectrometers. Even if one does not possess a computer, it is still possible to access a large library over telephone lines and via satellite.[35]

Obviously, it is highly uneconomic to search a large library of full spectra, so that some way must be found to abbreviate a spectrum while still retaining those features that differentiate it as far as possible from the spectra of other compounds. Various methods have been described for encoding spectra, and several algorithms have been developed for matching a spectrum against a library.

In the system developed by Biemann,[36,37] the spectra were abbreviated by taking the two most intense peaks in each 14 mass unit interval from m/e 9 upwards. A similarity index was then calculated so that the compounds with the best agreement could be printed out.

The similarity index requires a number of steps for its calculation, and must take into account the fact that the spectrum can be distorted if obtained in a g.c.m.s. system at incorrect scan speeds. (a) The ratio R of the intensity of each peak in the unknown to that in the possible match spectrum or vice versa is calculated, the smaller being divided by the larger so that the value of $R \leqslant 1$. R is given a value of zero if a peak is absent from either spectrum. (b) R is divided by the average R for all peaks that have an intensity greater than 10%. (c) Since it can be assumed that abundant peaks are more meaningful than weak peaks, the R values must be weighted to reflect this fact. The weighting factors are 12 for peaks $>10\%$, 4 for peaks between 10 and 1% and 1 for peaks less than 1%. (d) The average weighted ratio is calculated by adding the weighted ratios and dividing by the weighted number.

The similarity index S can then be determined as

$$S = \frac{\text{average weighted ratio}}{1 + \text{fraction of unmatched spectra}}$$

S will lie between 1 (perfect fit) and 0 for totally dissimilar spectra.

The 'Self Training Interpretive and Retrieval System' (STIRS) utilizes knowledge of mass spectral fragmentation processes.[38] The data which are compared between the unknown and a possible match spectrum are: (a) series of ions, e.g. $[C_nH_{2n-2}]^+$, $[C_nH_{2n+1}]^+$, etc.; (b) characteristic ions of high mass; (c) characteristic ions of low mass; (d) characteristic ions of medium mass; (e) losses of small primary neutral fragments; (f) losses of large primary neutral fragments; (g) secondary losses of neutral fragments from the most abundant odd mass ions; (h) secondary losses of neutral fragments from the most abundant even mass ions; (i) fingerprint ions.

A match factor is obtained by weighting the match factors of each of the classes (a) to (f) as follows: (a) − 1, (b) − 1, (c) − 2, (d) − 2, (e) − 4, (f) − 6.

These are then summed and divided by the weighted number (12) to give an overall match factor. Used on a file of 13 000 spectra, the system gave a high success rate to the identification of various structural features.

Another matching system has been described in which a reverse search is used.[39] The m/e values of the reference spectrum are searched in the unknown and the intensities compared. This has the advantage that extra peaks present in the unknown, perhaps from impurities, or because the spectrum has been obtained from an unresolved g.c. peak, will be ignored, and not cause a mismatch as is the case with some other systems.

The Clerc system[40] compares the 16 most heavily weighted features of the unknown spectrum with the library spectra, so that spectra which are very dissimilar are rejected rapidly. The probability of correct matching has been shown to improve when other spectral characteristics such as nuclear magnetic resonance and infrared are incorporated.[41]

Although in one sense large libraries of spectra, say around 50 000, are useful since there is a better chance that they contain a given compound, much time is wasted matching the spectra of compounds from a different area of research, which have a zero chance of matching. It is an advantage to construct sub-libraries of compounds likely to be encountered in a given line of research, for example steroids, drugs or normal urinary constituents.

REFERENCES

1. K. R. Jennings, *J. Chem. Phys.* **43**, 4176 (1965).
2. K. R. Ryan, L. W. Sieck and J. H. Futrell, *J. Chem. Phys.* **43**, 1832 (1965).
3. M. Barber, W. A. Wolstenholme and K. R. Jennings, *Nature (London)* **214**, 664 (1967).
4. E. M. Chait and W. B. Askew, *Org. Mass Spectrom.* **5**, 147 (1971).
5. S. Safe, O. Hutzinger and M. Cook, *Chem. Commun.* 260 (1972).
6. J. T. Watson, F. C. Falkner and B. J. Sweetman, *Biomed. Mass Spectrom.* **1**, 156 (1974).
7. C.-G. Hammar, B. Holmstedt and R. Ryhage, *Anal. Biochem.* **25**, 532 (1968).
8. D. Henneberg, *Z. Anal. Chem.* **170**, 369 (1959).
9. D. Henneberg, *Z. Anal. Chem.* **183**, 12 (1961).
10. J. B. Knight, *Finnigan Spectra* **1**, 1 (1971).
11. C. C. Sweeley, W. A. Elliot, I. Fries and R. Ryhage, *Anal. Chem.* **38**, 1549 (1966).
12. J. F. Holland, C. C. Sweeley, R. E. Thrush, R. E. Teets and M. A. Beiber, *Anal. Chem.* **45**, 308 (1973).
13. P. D. Klein, J. R. Haumann and W. J. Eisler, *Anal. Chem.* **44**, 490 (1972).
14. P. D. Klein, J. R. Haumann and W. J. Eisler, *J. Clin. Chem.* **17**, 735 (1971).
15. L. D. Gruenke and J. C. Craig, *Chem. Ind. (London)* 718 (1974).
16. L. D. Gruenke, J. C. Craig and D. M. Bier, *Biomed. Mass Spectrom.* **1**, 418 (1974).
17. K. Biemann, P. Bommer, D. M. Desiderio and W. J. McMurray, *Adv. Mass Spectrom.* **3**, 701 (1966).
18. W. J. McMurray, B. N. Green and S. R. Lipsky, *Anal. Chem.* **38**, 1194 (1966).
19. H. C. Bowen, E. Clayton, D. J. Shields and H. M. Stanier, *Adv. Mass Spectrom.* **4**, 257 (1968).
20. A. Carrick, *Int. J. Mass Spectrom. Ion Phys.* **2**, 333 (1969).
21. A. Carrick, *Adv. Mass Spectrom.* **5**, 330 (1971).
22. K. Biemann and P. V. Fennessey, *Chimia* **21**, 226 (1967).
23. K. Habfast, K. H. Maurer and G. Hoffman, American Society for Mass Spectrometry, 20th Annual Conference on Mass Spectrometry and Allied Topics, Dallas, Texas (1972). Proceedings, p. 414.
24. W. E. Reynolds, V. A. Bacon, J. C. Bridges, T. C. Coburn, B. Halpern, J. Lederberg, E. C. Levinthal, E. Steed and R. B. Tucker, *Anal. Chem.* **42**, 1122 (1970).
25. H. P. Hotz and V. L. Dagragno, American Society for Mass Spectrometry, 21st Annual Conference on Mass Spectrometry and Allied Topics, San Francisco, California (1973). Proceedings, p. 74.

26. D. E. Green and R. H. Hertel, Symposium on Analytical Advances in Clinical and Medicinal Chemistry, American Chemical Society, Chicago (1973).
27. K. Elkin, L. Pierron, U. G. Ahlbor, B. Holmstedt and J. E. Lindgren, *J. Chromatogr.* **81**, 47 (1973).
28. J. T. Watson, D. R. Pelster, B. J. Sweetman, J. C. Frolich and J. A. Oates, *Anal. Chem.* **45**, 2071 (1973).
29. J. H. Hengstmann, F. C. Falkner, J. T. Watson and J. A. Oates, *Anal. Chem.* **46**, 34 (1974).
30. L. Baczynskyj, D. J. Buchamp, J. F. Zieserl Jr and J. A. Axen, *Anal. Chem.* **45**, 479 (1973).
31. R. A. Hites and K. Biemann, *Anal. Chem.* **42**, 855 (1971).
32. J. E. Biller and K. Biemann, American Society for Mass Spectrometry, 22nd Conference on Mass Spectrometry and Allied Topics, Philadelphia, Pennsylvania (1974). Proceedings, p. 430.
33. J. E. Biller and K. Biemann, *Anal. Lett.* **7**, 515 (1974).
34. R. G. Dromey, M. J. Stefik, T. C. Rindfleisch and A. M. Duffield, *Anal. Chem.* **48**, 1368 (1976).
35. S. R. Heller, H. M. Fales and G. W. A. Milne, *J. Chem. Doc.* **13**, 130 (1973).
36. H. S. Hertz, R. Hites and K. Biemann, *Anal. Chem.* **43**, 681 (1971).
37. H. S. Hertz, D. A. Evans and K. Biemann, *Org. Mass Spectrom.* **4**, 452 (1970).
38. K.-S. Kwok, R. Venkataraghavan and F. W. McLafferty, *Am. Chem. Soc.* **95**, 4185 (1973).
39. F. P. Abramson, American Society for Mass Spectrometry, 21st Annual Conference on Mass Spectrometry and Allied Topics, San Francisco, California (1973). Proceedings, p. 76.
40. J. T. Clerc, F. Erni, C. Jost, T. Meili, D. Nageli and R. Schwarzenbach, *Z. Anal. Chem.* **264**, 192 (1973).
41. F. Erni and J. T. Clerc, *Helv. Chim. Acta* **55**, 489 (1972).

CHAPTER 3

SOURCES OF ERROR

In a quantitative mass spectrometric analysis, the overall error for the determination will result from the cumulative effect of the errors introduced by different parts of the system. If improvements in the analytical procedure are to be carried out, it is essential that the various sources of error in the complete system are recognized, along with their relative importance. A reasonably logical division can be made between errors attributable to the mass spectrometer and data system, if the latter is employed, and errors due to sample handling and introduction.

MASS SPECTROMETRIC AND DATA SYSTEM ERRORS

Mass measurement

It is a widely held view that the accuracy of mass measurement is directly dependent upon the resolving power of the instrument. This seemed to be confirmed by work carried out by McLafferty,[1] who compared the mass measuring errors for the ions of n-decane at resolving powers of 2000 and 10 000. At 2000 resolution the average error distribution for 18 spectra (68 % confidence level) was 38.5 ppm, while the corresponding figure at 10 000 resolving power was 59.1 ppm. The values were shown to be significantly different, so that the use of higher resolving powers yielded higher accuracy. This conclusion was disputed by Burlingame, who carried out accurate mass measurements at resolving powers of 2500, 5000 and 10 000, using a scan rate of 16 s per decade.[2] The errors were a factor of 10 lower than those of McLafferty, and there was no significant difference between the errors at the three resolving powers.

The feasibility of carrying out accurate mass measurements with an error of a few parts per million at low resolving power, provided that the ions concerned are not unresolved multiplets, is now accepted. Thus, by peak matching, a molecular ion, which is obviously a singlet unless it has the same nominal mass as a peak in the reference compound, can be measured against a reference peak of higher m/e value, since the latter cannot have any compound peaks forming a multiplet with it. It is less safe to rely on mass measurements carried out on any other peak in the spectrum, except perhaps $[M - 15]^+$ and $[M - 18]^{+}$, unless a computer evaluation of the peak width is used to determine its homogeneity. The normal peak width for a singlet can be determined by a run on PFK by itself. The reason that low resolution measurements are feasible is that mass measurement accuracy is a function of the product of resolution and sensitivity[2] which is essentially constant. This has been supported by measurements carried out on the A.E.I. MS-30 double beam instrument in which sample and reference PFK are kept separate, so that the vast majority of peaks are singlets. The relevant data are given in Table 3.1.[3]

TABLE 3.1
Mass measurement errors in two modes of operation of an MS-30 mass spectrometer

Elemental composition Dichlorobenzene $[M]^{+}$	Relative intensity	Calculated mass	Error (m.m.u.) Mode A[a]	Error (m.m.u.) Mode B[b]
$C_6H_4{}^{37}Cl_2$	8.4	149.9630	−0.5	Not observed
$C_6H_4{}^{35}Cl{}^{37}Cl$	57.2	147.9660	0.8	1.4
$^{13}C^{12}C_5H_4{}^{35}Cl_2$	8.0	146.9723	−1.2	3.1
$C_6H_4{}^{35}Cl_2$	100.0	145.9689	−1.2	2.4

[a] Mode A: 50 ng on g.c. column, 1500 resolution, double beam.
[b] Mode B: 250 ng on g.c. column, 10 000 resolution, single beam.

The mass measurement accuracy at 1500 resolving power on a 50-ng sample is superior to that at 10 000 resolving power on a 250-ng sample, thus confirming Burlingame's proposition. Further evidence to support this has also been put forward for a single beam instrument, where accuracies of the order of a few parts per million have been obtained.[4] From this it follows that in those frequent cases where elemental compositions are required only on molecular ions, the sensitivity advantage of working at low resolving power should be utilized. As a general rule, one should work at the lowest resolving power commensurate with resolving the peaks of interest.

At low resolving power with PFK as reference, most of the multiplets present in the spectrum are due to the coincidence of sample and reference peaks of the same nominal mass. The use of PFK as an *external* calibrant for setting up the mass scale, followed by iodoform as an internal calibrant, can extend the usefulness of the low resolution mass measurement method to all those sample peaks which are themselves singlets. Mass measurement accuracies of 16 ppm at a resolving power of 1200 have been reported using this method.[5]

With a data acquisition system, the accuracy of mass measurement is vastly more dependent upon the setting up of the mass spectrometer than on the data system itself. It has been reported[6] that the average mass errors for major ions lie between ± 3.1 and ± 5.7 ppm when the mass spectrometer is properly adjusted, while another report quotes an average error of better than ± 10 ppm.[7] The importance of mechanical vibration is shown by the fact that its removal improved the mass measurement accuracy to ± 3 ppm. Installation engineers for high resolution instruments usually check sites for vibration, and where necessary suspend a concrete raft on compressed air mountings to isolate the instrument. Mechanical pumps can be a major source of vibration, and steps should be taken to isolate these by correct mounting on rubber feet and the use of damped vacuum lines. The consensus of opinion is that a mass measurement accuracy of better than ± 10 ppm should be attainable at a scan speed of 10 s per decade on about 100 ng of sample through a g.c. inlet. Rather more sample is required to achieve the same performance with a direct inlet.

As the sample level falls, the number of ions in a peak falls, and the accuracy of mass measurement becomes limited by ion statistics. An examination of the theoretical limit of mass measurement accuracy has been carried out by Campbell and Halliday[8] and their results are presented in Table 3.2.

TABLE 3.2
Error (ppm) expected in measurement of 100 ppm peak

Number of ions in peak	Confidence level		
	$\pm 64\%$	$\pm 95\%$	$\pm 99\%$
1	20.4 ppm	38.8 ppm	45.0 ppm
2	14.4	28.0	35.3
3	11.8	22.9	29.4
4	10.2	19.9	25.7
10	6.45	12.7	16.5
30	3.7	7.3	9.6
100	2.0	4.0	5.25
300	1.2	2.3	3.0
1000	0.065	1.3	1.7

McLafferty[1] has identified the major sources of error in high resolution data acquisition systems. Mass spectrometric errors are due to mechanical vibration and electronic noise, both of which degrade the shapes of the peaks. In the analogue system and interface the bandwidth of the signal is important. However, in a well-maintained instrument, ion statistics will be the most important source of error in the signal which arrives at the a.d.c.

The a.d.c. can impose further errors due to the rate of digitization and the bit number of the a.d.c. The digitization rate must be low enough to avoid unnecessary calculation time by the computer, but high enough to define the peak shape. The distortion which can occur with a low digitization rate can be seen in

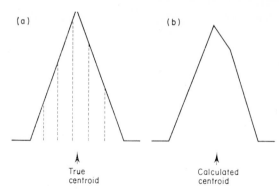

Fig. 3.1.   The distortion introduced by a low digitization rate: (a) original peak and position of true centroid. This corresponds with the apex of the peak since the latter is symmetrical; (b) peak reconstructed from the five points sampled on the original peak. The calculated centroid is in error due to the fact that a sampling point does not coincide with the peak maximum.

Fig. 3.1. Most data acquisition systems employ a calculation of the peak centroid, as follows:

$$\text{centroid} = \frac{\Sigma V_n X_n}{\Sigma V_n}$$

where V_n is the voltage sampled at time X_n.

The centroid positions of the peaks, as time values, are then used for calculation of accurate mass values, based on interpolation between the times of the reference peaks whose accurate masses are known (Chapter 1).

This means that a considerable error, whose value is dependent upon where sampling commences over the profile of the peak, is involved when only a small number of samples is taken across the peak. This error naturally decreases as the sampling rate increases.

The sampling rate necessary to yield N samples in a peak during an exponential scan is given by

$$S = 2.3 \times 10^{-3} \frac{NR}{t_{10}}$$

where R = resolution of the mass spectrometer, t_{10} = scan time per decade in mass, and S is in kilosamples per second.

A reasonable compromise between computer time and centroiding error is obtained when about 15 to 20 samples are taken across the peak. This number of samples defines its shape adequately.

Since the calculation of the centroid additionally involves an intensity, or rather a voltage measurement, mass measurement error can also be introduced by the analogue-to-digital conversion process. An error of 1 bit in a 10-bit digitizer corresponds to one part in 1024, i.e. about 0.1% of the maximum voltage. Corresponding figures are about 0.05% for 11 bits and 0.025% for

12 bits. However, the calculation of the centroid is a weighting process in which the larger voltages are more significant than the smaller ones, so that unless a serious defect is present in the a.d.c., the overall effect on the position of the centroid is small.

Intensity measurement

Effect of changes in source parameters

Whichever method of sample introduction into the mass spectrometer is used, whether it is direct inlet, hot or cold reservoir or molecular separator, the mass

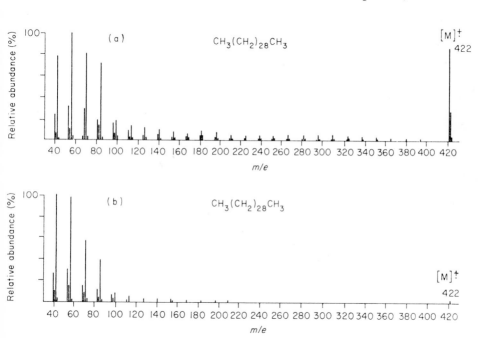

Fig. 3.2. Effect of temperature on ion abundance. The mass spectra of triacontane at source temperatures of (a) 70 °C and (b) 340 °C are compared. (Reproduced with permission from Ref. 9.)

spectrum obtained must be reproducible on the same instrument over a period of time if reproducible analyses are to be obtained. A variation of a few percent in relative intensities is usual over a period of a few days, but this should not be more than about 1% over the course of a few hours. Factors which affect the reproducibility of a spectrum include the temperature of the source and inlet system, the state of cleanliness of the source, the condition of the filament, the electron beam energy, the trap current and the repeller voltage. Before serious quantitative work is carried out, it is essential that the source is tuned for maximum stability.

Temperature. The effect of temperature on relative ion abundance can be considerable, as has been shown by Spiteller.[9,10] The spectra of triacontane run at 70 °C and 340 °C (Fig. 3.2) show a tremendous difference in the relative abundance of the molecular ion. This is 86% of the intensity of the ion at m/e 57 at 70 °C, but only 1% of the same ion at 340 °C.

Admittedly these are extremes of temperature, but even a change in temperature of from 175 to 225 °C leads to a five-fold decrease in the molecular ion intensity in a branched chain hydrocarbon such as 2,2,3-trimethylpentane.[11] Where compounds themselves are thermally unstable, the situation can be even more difficult. For example, *N*-oxides are particularly sensitive to thermal degradation, although if care is taken, molecular ions can often be observed. Thus, it has been reported that 4-methylmorpholine-4-oxide (**1**) gives a molecular

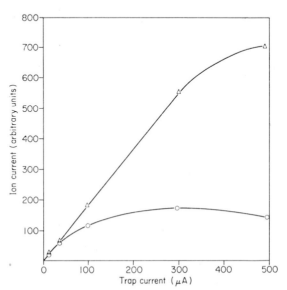

1

ion in a direct inlet system when the probe is unheated.[12] The intensity was markedly reduced at a probe temperature of 50 °C, while at 110 °C the molecular ion was not detectable. The interesting point in these experiments is that the source temperature was kept constant at 250 °C. This indicates that in the case of thermal decomposition at least, the probe temperature was of far greater

Fig. 3.3. Variation of ion abundance with trap current. $\triangle$: the $[CF_3]^+$ ion from PFK; $\bigcirc$: the $[C_2F_5]^+$ ion from PFK.

importance than the source temperature in influencing the extent of thermal decomposition, and hence ion intensities. This has implications for quantification using a direct inlet probe, since frequently a programmed probe temperature is used. A failure to reproduce the programme exactly on subsequent runs may lead to a variation in ion intensity even though the source temperature is held constant. However, the design of the source itself will also have a bearing on the extent to which source temperature causes thermal decomposition.

Filament. The condition of the filament can have an effect on ion intensities since the emission may change over the short term. The act of introducing the sample alters the work function of the filament, thus altering the electron emission, which leads to a change in source temperature. A change in electron current can have a profound effect on relative ion abundances. Some appreciation of the effect can be obtained from Fig. 3.3, where the intensity of the $[CF_3]^+$ and $[C_2F_5]^+$ ions from PFK are plotted. While a small change in trap current has a minimal effect on the intensity of the $[C_2F_5]^+$ ion, it has a marked effect on that of the $[CF_3]^+$ ion, especially for a trap current of 30 μA.

Electron voltage. A beam energy of 70 eV is usually accepted as the standard for recording mass spectra, mainly because this point lies on a fairly horizontal portion of the ionization efficiency curve. A typical curve would appear as in Fig. 3.4.

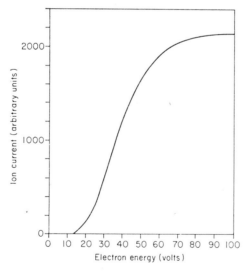

Fig. 3.4. A typical ionization efficiency curve for an organic compound.

In the case of a fragment ion, a large number of competing reactions are involved in deciding its final abundance. Since these reactions require varying amounts of energy, their relative importance changes as the beam energy is decreased. Obviously, if the instrument is operated at 70 eV, small changes in beam energy have little effect on the relative intensities of ions in the spectrum. Many applications require low voltage determinations, however, in order to reduce contributions from interfering substances in the case of complex mixtures. It has to be accepted that such determinations may well suffer from greater error due to operation at a part of the ion current–electron beam energy curve which has a steep slope.

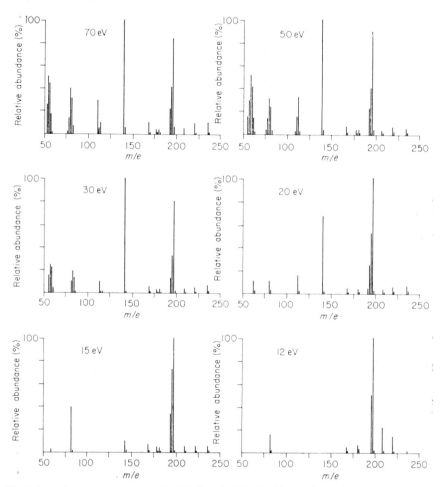

Fig. 3.5. The mass spectrum of *N,N*-dimethylallobarbitone through a g.c.m.s. system, taken at a number of beam energies from 70 down to 12 eV.

The effect of decreasing the beam energy on peak ratios can be seen from Fig. 3.5, where mass spectra of *N,N*-dimethylallobarbitone have been recorded at beam energies of 70, 50, 30, 20, 15 and 12 eV. Although the ratio of *m/e* 195 to 138 remains fairly constant in dropping from 70 to 50 eV, the intensities of lower mass peaks exhibit an obvious change. The change between successive spectra becomes even more pronounced on dropping even further.

The overall impression gained from these spectra is that ions of lower m/e value are affected by a drop in beam energy to a greater extent than those at higher values. Thus, it is more advantageous to choose ions of high m/e value for quantitative measurements. As discussed later (p. 128), the effect of interfering substances in complex mixtures also decreases as higher m/e values are monitored.

Repeller voltage. Typical ion repeller voltage–ion current curves are shown in Fig. 3.6. Two maxima are apparent; the first, more intense maximum occurring

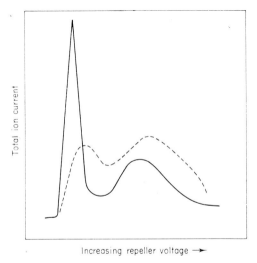

Increasing repeller voltage →

Fig. 3.6. Relationship between total ion current and repeller voltage. Solid line: source tuned for maximum ion current before changing the repeller voltage. Broken line: source tuned at second maximum before changing repeller voltage.

at a lower ion repeller voltage. The shape of the curves depends upon the way in which the source has been tuned. The solid curve with the higher maximum was obtained by maximizing all the source controls and then changing the repeller voltage. The dotted curve was obtained by adjusting the repeller voltage to the position of the second maximum and resetting the other source controls before changing the repeller voltage. Obviously, if the greatest sensitivity is required, the source should be tuned in the first way with the repeller voltage set for the first maximum. However, for quantitative work where sensitivity is not a prime requirement, greater stability is obtained in tuning the source to the second maximum as above. A greater dynamic range of sample levels can be accommodated at the second maximum since it increases linearly with sample level well past the point at which the first maximum saturates. Occasionally, with a particularly high sample pressure, and a repeller set at the first maximum, it has been noticed that the ion current drops suddenly to a fraction of its previous value, although the reason for this is not apparent.

Magnet drift

The problem of magnet stability is not as serious on instruments manufactured in the last few years as on older instruments, which were generally designed for scanning a spectrum rather than remaining focused at a particular m/e value. One instrument has been shown to drift as much as 0.4 of a mass unit over a period of 200 min, as shown in Fig. 3.7.

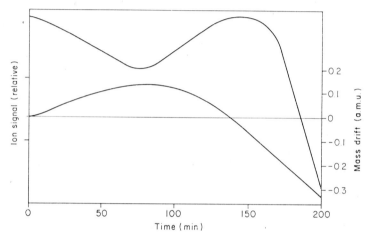

Fig. 3.7.   Magnet drift characteristics of the LKB 9000S at m/e 505, cold magnet at zero time. Upper trace: ion signal; lower trace: mass drift in a.m.u. (Reproduced with permission from G. S. King, D. Talbert and J. Mangez, *Biomed. Mass Spectrom.* **4**, 62 (1977).)

The effect of this drift on the intensity of an ion in focus at the start of the experiment is also shown in the figure. The ion drifts out of focus in one direction and then back again before remaining out of focus for a long period. As long as one is aware of this problem, it can be compensated for by frequent checking of the magnet tuning throughout the course of an analysis.

Recording and amplifying system

In addition to errors caused by fluctuating source parameters, there are errors in the intensities of peaks arriving at the recorder or data system due to noise arising from the various amplification stages and noise due to ion statistics. It is important that the instrument is checked frequently so that noise from the multiplier and amplifiers should remain within specification. In a well-maintained instrument the reproducibility of intensity measurements should be limited by ion statistics rather than electronic noise. If N is the average number of ions in a peak, the theoretical percentage standard deviation, i.e. that due solely to ion statistics, will be $100/\sqrt{N}$. Table 3.3[13,14] shows that a reasonable correlation

TABLE 3.3
Correlation between theoretical ($100/\sqrt{N}$) and observed standard deviation for peaks in a mixture of tribenzylamine and PFK[13,14]

Peak	Average number of ions in peak	$100/\sqrt{N}$	Observed percentage standard deviation
$[C_7F_{11}]^+$	140	8.5	10.7
$[^{13}CC_{20}H_{21}N]^+$	50	14.1	15.3
$[C_{21}H_{21}N]^+$	230	6.6	8.3
$[C_{21}H_{20}N]^+$	50	14.1	12.1
$[C_8F_{10}]^+$	15	25.8	25.0
$[^{13}CC_5F_{11}]^+$	15	25.8	38.5
$[C_6F_{11}]^+$	230	6.6	3.7

exists between the theoretical and observed percentage standard deviation for peak areas calculated for some peaks in a mixture of tribenzylamine and PFK.

As discussed later (p. 54), quantitative applications of mass spectrometry primarily involve measurement not of the absolute intensity of a peak but rather the ratio between two or more peaks. Since Table 3.3 shows that ion statistics are the major source of error in the measurement of peak intensities, it is possible to calculate the error involved in measuring the ratio of the intensities of two peaks for given numbers of ions in each peak.[13] The results of this calculation are shown in Fig. 3.8.

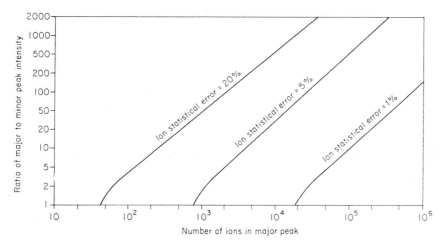

Fig. 3.8. The ion statistical error involved in measuring the ratio of two peaks. (Reproduced with permission from Ref. 13.)

The important points illustrated by the figure are, first, that the error decreases as the number of ions in the major peak increases, and, second, that the error in the ratio decreases as the ratio approaches unity.

Thresholding

In measuring the ratio of two peak intensities, the position of the threshold can have a serious effect on the peak ratio which is actually observed. Thresholding becomes particularly important in data acquisition systems, as most systems are limited in the number of peaks they can accommodate in a scan. Since noise peaks can be processed as mass spectral peaks, it is usual to set a hardware threshold to sit just above the noise. Thresholds of up to 10 mV are common. The visual effect of a threshold on the spectrum obtained can be seen in Fig. 3.9,

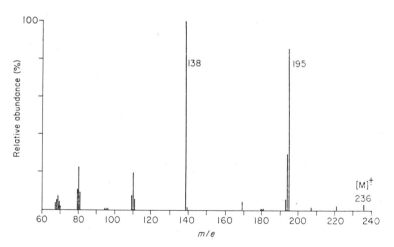

Fig. 3.9.   The mass spectrum of dimethylallobarbitone run with a high threshold on the data acquisition system.

where the isotope peaks, which range usually from a few percent to 10% for a compound containing nine carbon atoms, are either absent, or unusually small. This gives the spectrum a rather odd appearance.

The effect on the ratio of two peaks of various thresholds is given in Fig. 3.10(a), where the effect of 1, 5 and 10 mV thresholds is given on the measurement of a peak ratio of 10:1 as the theoretical voltage of the signal due to the smaller peak is reduced. A 10-mV threshold on a pair of peaks of theoretical voltage 200 mV and 2 V results in a 4% error in the ratio. Since the major error is caused in the smaller of the two peaks, not surprisingly, the curves for a larger peak ratio, such as 100:1, are virtually the same, as shown in Fig. 3.10(b). As

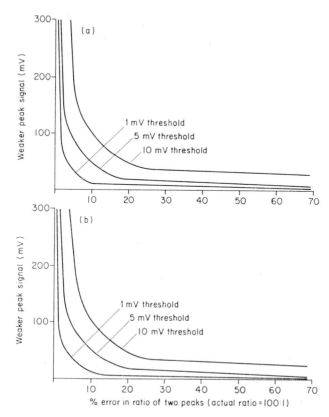

Fig. 3.10. Error in the measurement of the ratio of two peak intensities versus the weaker peak intensity for thresholds of 1, 5 and 10 mV: (a) for a peak ratio of 10:1; (b) for a peak ratio of 100:1.

the ratio approaches 1, the error decreases until it becomes zero when the two peaks are of equal intensity.

Peak heights and areas

There is still much controversy as to whether peak heights or areas should be used for intensity measurements. An early paper[15] noted that either peak heights or areas could be used with about the same reliability. Current thinking is that peak heights are more reliable for weak peaks. If the peak is well resolved from other components, it is relatively easy to decide upon a base line level, so that there is really only one uncertain measurement, that of the height of the top of the peak. However, in calculating peak areas, decisions must be made about when the peak commences and ends. In the absence of a data system, the ease of measuring a peak height rather than its area is also a consideration. Unless there is good reason for concluding that the peak width or shape is changing over successive runs, peak height measurements should prove to be perfectly adequate.

The most commonly used method of determining a peak height at present is to measure it on chart paper with the aid of a ruler. By this means it is normally possible to measure a peak to the nearest 0.5 mm. This leads to a considerable error when peaks are only a few centimetres high. In order to reduce this measuring error, the recorder sensitivity should be adjusted where possible to bring the peak to about three-quarters of full-scale deflection. This will accommodate any reasonable fluctuations in sample size over consecutive runs so that no off-scale responses are obtained. If it is felt necessary to reduce the variance of ruler measurements by remeasuring and taking a mean value, it should be borne in mind that there is an unconscious tendency to remember the previous reading. It is more meaningful to have the remeasurements carried out by another person.

The error introduced into intensity measurements by analogue-to-digital conversion has been mentioned already in connection with mass measurement accuracy. A 1-bit error in a 12-bit a.d.c. amounts to about 0.025% of the maximum voltage. With a 12-bit converter, the error introduced by digital conversion will be less than the error in the signal due to ion statistics and electronic noise.[14]

SAMPLE HANDLING AND INTRODUCTION

Propagation of errors

A typical sequence for the analysis of a sample by mass spectrometry would be that shown in Fig. 3.11, involving seven distinct stages.

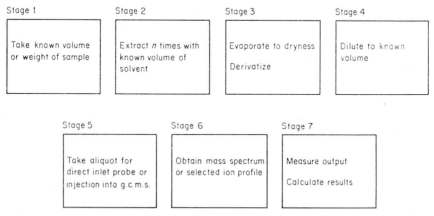

Stage 1	Stage 2	Stage 3	Stage 4
Take known volume or weight of sample	Extract *n* times with known volume of solvent	Evaporate to dryness Derivatize	Dilute to known volume

Stage 5	Stage 6	Stage 7
Take aliquot for direct inlet probe or injection into g.c.m.s.	Obtain mass spectrum or selected ion profile	Measure output Calculate results

Fig. 3.11.   A typical sequence for a quantitative analysis by mass spectrometry.

Errors are inherent in every one of the seven stages of this sequence, some of these being more serious than others. If the result obtained for the analysis at the end of the seven stages outlined above is R, then R can be expressed as the product of the true result r and a set of variables $x_1 \ldots x_7$ representing the seven stages. Thus, $R = rx_1x_2x_3x_4x_5x_6x_7$.

If there were no errors in any of the stages, each x would be equal to unity, and obviously the result obtained would be the true result. In real life, such a state of affairs does not exist. For example, the volume of sample taken in stage 1 might be 0.5% less than the recorded value, so that x_1 would be 0.995. If in stage 4 the sample was taken up to 101 μl instead of 100 μl, x_4 would be 0.990. Similar reasoning can be applied throughout the complete sequence.

Each variable x_n is associated with the coefficient of variation (c.v.) of the particular operation it represents (see Appendix). The c.v. of a product R ($= rx_1x_2x_3x_4x_5x_6x_7$) is given by

$$\text{c.v.}\,(R) \approx \sqrt{\{[\text{c.v.}\,(r)]^2 + [\text{c.v.}\,(x_1)]^2 + \cdots + [\text{c.v.}\,(x_7)]^2\}}$$

However, c.v. $(r) = 0$ since r is the true result, and therefore

$$\text{c.v.}\,(R) \approx \sqrt{\{[\text{c.v.}\,(x_1)]^2 + \cdots + [\text{c.v.}\,(x_7)]^2\}}$$

Thus, the coefficient of variation of the determined result is approximately equal to the square root of the sums of the squares of the individual coefficients of variation. Some typical reasons for large coefficients of variation at each of the seven stages shown in Fig. 3.11 are given in Fig. 3.12.

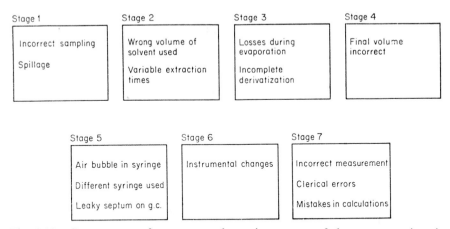

Fig. 3.12. Some reasons for errors at the various stages of the sequence given in Fig. 3.11.

Potentially the most serious errors are caused by spillage at any stage and spitting of sample during evaporation to dryness. These can cause exceedingly large coefficients of variation for the appropriate stage of the analysis. In the absence of these gross errors, probably the two most serious sources of error are in the sampling at stage 1 and the injection of sample onto the gas chromatograph or mass spectrometer probe.

In general, the type of problem encountered in quantitative mass spectrometry falls into two categories. The first is determination of the amount of a component in a fairly simple mixture where all the other components are known. It is necessary to establish the ratio of the component to one or more of the other components. The errors outlined in the above paragraph become of minimum importance, since the ratio of the components remains unchanged in spite of spillage, incorrect sampling, gravimetric, volumetric or poor g.c. injection technique. It will still be important to take care over extraction procedures, since differential extraction of some components may not be reproducible, leading to fluctuating component ratios. Derivatization must be controlled carefully in order to ensure that all components are completely derivatized. Usually, the largest source of error in problems of this type will be that involved in making up the standard mixtures from which the calibration graph will be constructed. The errors inherent in the mass spectrometry will be much less significant. An example of this type of problem is the determination of an impurity in a sample, where the impurity is to be expressed as a percentage of the major component.

A simple experiment serves to show that the errors in making up solutions are usually greater than those involved in the mass spectrometric determination. Four solutions were made up by weighing out 1 mg of dimethylallobarbitone and dissolving this in 100 ml of methanol. Thus 1 μl of these solutions contained a nominal 10 ng. For each of the four solutions four injections of 1 μl were made into the g.c.m.s. while monitoring the response at m/e 195, with output on a pen recorder. The results obtained are given in Table 3.4. The errors can be estimated

TABLE 3.4
Responses for four 1 μl injections of each of four solutions of dimethylallobarbitone, nominal concentration 10 ng μl^{-1}

Solution	Responses at m/e 195 (in cm on chart)			
A	8.40	8.04	8.25	8.11
B	6.40	6.06	6.25	6.13
C	7.04	7.14	7.22	7.10
D	7.82	7.78	7.90	7.32

TABLE 3.5
Analysis of variance of data from Table 4

Source of variance	Sums of squares	Degrees of freedom	Mean squares	Variance given by mean squares
Between solutions	8.76944	3	2.9231	
Within solutions	0.37500	12	0.03042	$v_{\text{g.c.m.s.}}$
Total	9.13444	15	0.60866	v_{total}

by calculating the variance for the mass spectrometric determination and the variance for the whole procedure. The results are given in Table 3.5.

The variance of the g.c.m.s. determination, $v_{g.c.m.s.}$, is the sum of the variance of the injections onto the gas chromatograph, the variance of the mass spectrometric response, and the variance of the measurement of the peak heights from the chart paper. For the purposes of the present experiment, it is not necessary to break down $v_{g.c.m.s.}$ into its components. $v_{g.c.m.s.}$ can be calculated from replicate injections of a single sample by the formula

$$v_{g.c.m.s.} = \frac{(\text{response} - \text{mean response})^2}{n - 1}$$

The responses are the peak heights measured from the chart paper.

Variance calculations always consist of a 'sums of squares' term divided by a 'degrees of freedom' term. In the simplest case as given above, the number of determinations is n, but the number of degrees of freedom is $n - 1$, since the mean is used in the calculation, and thus there are only $n - 1$ independent results. In order to use all the responses given in Table 3.4 for the calculation of $v_{g.c.m.s.}$, a slightly different approach is required. The sums of squares term must be calculated by adding the squares of the differences between the four results for a sample and the mean of that sample, repeated for each of the four samples. This gives 16 terms, which total 0.37500, as shown for the 'within solutions' row in Table 3.5. The calculation uses 4 means and 16 results, so this leaves 12 degrees of freedom. Dividing the sums of squares, 0.37500, by degrees of freedom, 12, yields $v_{g.c.m.s.} = 0.03042$.

The variance of the gravimetric and volumetric procedures in making up the solutions, v_s, cannot be obtained directly. This is because any mass spectrometric response already includes $v_{g.c.m.s.}$ (see Appendix). It can be obtained by difference, however. The total variance of the whole series of determinations, v_{total}, is given by the usual formula from the 16 responses and the grand mean of these. Table 3.5 shows this value to be 0.60866. The solution preparation variance, v_s, is now given by

$$v_s = v_{total} - v_{g.c.m.s.} = 0.60866 - 0.03042 = 0.57824$$

The F test can be used to test for a significant difference between v_s and $v_{g.c.m.s.}$ (see Appendix). In the present case the ratio $v_s/v_{g.c.m.s.}$ is 19, and statistical tables show this value (F) to be significant at the 0.95 probability level.

The error involved in making up the solutions in this experiment is obviously many times greater than the error of the g.c.m.s. determination. This fact, frequently ignored by many workers, has serious repercussions on the construction of calibration curves, especially from isotope dilution data, and is discussed in Chapter 4.

It is, of course, dangerous to generalize from such results. If larger quantities are weighted out, say 100 mg, in order to make up solutions, the error should be less for this particular operation. However, if several dilution stages are intro-duced, the error will be much greater. Again, the error in injecting 1 μl from a 1-μl syringe will be less than if 0.1 μl is injected from the same syringe, while 1 μl injected from a 10-μl syringe will be subject to a greater error. Poor injection technique coupled with good gravimetric procedure could well cause a situation where the error of the g.c.m.s. determination is much greater than the error in making up the solution. All factors such as these should be taken into account in designing an assay.

Where calibration lines are constructed by spiking aliquots of urine, plasma, etc., with compound and internal standard, followed by an extraction and derivatization, the precision of such determinations is always much worse than that obtained on experiments with simple solutions. This can be due to factors such as protein binding and the presence of interfering compounds at differing levels in the various aliquots of biological sample which have been spiked.

The other type of problem is that widely encountered in medicine and biochemistry, in which the absolute amount of a few components in a complex extract has to be determined. The complexity of the extract is such that the identity of many components is unknown. One approach, used by many workers in the past, is to determine the response of the mass spectrometer to a known amount of the compound of interest, and then to measure the response to the mixture being analysed. In such a situation, besides the errors in making up solutions, as discussed above, there are problems of spillage of solution, extrac-tion losses and losses on injection due to poor injection technique. Many of these difficulties are overcome by the use of an internal standard, so that the problem then becomes similar to the first type discussed, where the ratio of responses rather than absolute responses is of interest.

Internal standards

While internal standards are considered to be the universal panacea, there can be occasions where their use is disadvantageous. It has been noted already that the making up of solutions can lead to large errors. The use of an internal standard therefore of necessity introduces the error inherent in making up a solution of it, plus the error in taking aliquots to add to the solution under study. One should be certain that the reduction of errors through the various stages of analysis exceeds the error introduced by the addition of the standard. The later in a procedure the internal standard is introduced, the smaller the improvement which will be obtained, since there are fewer stages available for the reduction of sample loss errors.

If we take the process of sample injection and mass spectrometer response measurement, and assume that the mass spectrometer is at least in average condition, it can be shown that the advantage or disadvantage of an internal standard depends entirely on whether the injection technique is good or bad. To one of the solutions used in the last experiment, a quantity of dimethylquinal-barbitone was added. Since the latter, like dimethylallobarbitone, has an intense fragment ion at m/e 195, the amount added was adjusted to give an approximately similar response at m/e 195 for the two components. This mixture of compounds offers the possibility of measuring the variance of the response for one compound, which will reflect the variance in injection volume and mass spectrometer response, and of measuring the variance in the ratio of the two responses. This tests whether the use of one compound as internal standard for the other compound is an improvement over repeated injections of just one compound. The responses for these two compounds, and the ratio of the responses, are given in Table 3.6.

TABLE 3.6
Responses given for four injections of a mixture of dimethylallobarbitone and dimethylquinalbarbitone by an operator with good injection technique

Response at m/e 195 (in cm on chart)		Ratio of responses
Dimethylallobarbitone	Dimethylquinalbarbitone	
8.40	8.35	1.0059
7.65	8.33	0.9183
7.91	7.65	1.0299
7.68	7.94	0.9672
c.v. 4.38%	4.01%	4.97%

The coefficients of variation of the absolute responses ($\sim 4\%$) are such that the injections can be considered to be good. The coefficient of variation of the ratio is slightly worse, so that in a case such as this, as far as the g.c.m.s. system is concerned, the use of an internal standard is superfluous. An internal standard can be justified in this situation only if there are losses in the manipulations prior to injection.

It is interesting to see the picture that emerges when the injection technique is poor. A second operator carried out another four injections of the same solutions as above, this operator having a rather slipshod technique. The results are shown in Table 3.7.

TABLE 3.7
Responses for four injections of a mixture of dimethylallobarbitone and dimethylquinalbarbitone by an operator with poor injection technique

Response at m/e 195 (in cm on chart)		Ratio of responses
Dimethylallobarbitone	Dimethylquinalbarbitone	
5.80	6.42	0.9034
8.55	9.80	0.8724
7.69	8.15	0.9436
6.3	7.4	0.8514
c.v. 17.8%	31.1%	4.52%

The coefficients of variance for the absolute responses are so bad as to make the results almost worthless, yet the situation is now retrieved by considering one compound to be an internal standard for the other. The same overall effect would be obtained for any operation prior to the g.c.m.s. determination where the reproducibility was poor. All this leads to the conclusion that the great advantage of an internal standard is in its compensating effect for poor technique, and the better the technique, the less likely it is that an internal standard will be needed.

At present, internal standards can be divided into three types: A—a stable isotope labelled analogue of the compound to be measured, B—a homologous compound which yields a fairly intense ion in common to the compound to be measured, and C—a compound which yields a different ion, but which has similar solvent extraction and gas chromatographic properties to the compound under determination. The advantages and disadvantages of each of these types depend on the particular application, and are discussed later in the relevant chapters. Types A and C can be used in both chromatographic and non-chromatographic inlet systems, while type B is of use only in chromatographic inlets.

Equilibration of internal standards. A problem can arise frequently with protein binding when internal standards are added to various biological fluids. Variable ratios of compound of interest to internal standard can result upon extraction from such fluids if insufficient time is allowed for the internal standard to equilibrate. The procedure in this laboratory is to add internal standards to plasma, etc., the night before and carry out the extractions the following morning. It is probably more economical to carry out some experiments with different equilibration times in order to ascertain the minimum necessary for the particular internal standard being used.

REFERENCES

1. R. J. Klimowski, R. Venkataraghavan and F. W. McLafferty, *Org. Mass Spectrom.* **4**, 17 (1970).
2. D. H. Smith, R. W. Olsen, E. C. Walls and A. L. Burlingame, *Anal. Chem.* **43**, 1796 (1971).
3. M. L. Aspinal, J. R. Chapman, K. R. Compson, D. Hazelby and A. Riddoch, American Society for Mass Spectrometry, 21st Annual Conference on Mass Spectrometry and Allied Topics, San Francisco, California (1973). Proceedings, p. 471.
4. C.-G. Hammar and R. Hessling, *Anal. Chem.* **43**, 299 (1971).
5. W. F. Haddon and H. C. Lukens, American Society for Mass Spectrometry, 22nd Annual Conference on Mass Spectrometry and Allied Topics, Philadelphia, Pennsylvania (1974). Proceedings, p. 436.
6. R. S. Gohlke, G. P. Happ, D. P. Maier and D. W. Stewart, *Anal. Chem.* **44**, 1484 (1972).
7. R. M. Hilmer and J. W. Taylor, *Anal. Chem.* **45**, 1031 (1973).

8. A. J. Campbell and J. S. Halliday, American Society for Mass Spectrometry, 13th Annual Conference on Mass Spectrometry and Allied Topics, St Louis, Missouri (1965). Proceedings, p. 200.
9. M. Spiteller-Friedmann, S. Eggers and G. Spiteller, *Monatsh. Chem.* **95**, 1740 (1964).
10. M. Spiteller-Friedmann and G. Spiteller, *Chem. Ber.* **100**, 79 (1967).
11. J. H. Beynon, *Mass Spectrometry and its Application to Organic Chemistry*, Elsevier, Amsterdam, 1960, p. 429.
12. J. K. Faulkner and K. J. A. Smith, *J. Pharm. Pharmacol.* **26**, 473 (1974).
13. J. S. Halliday, in *Advances in Mass Spectrometry*, Vol. 4 (E. Kendrick, editor),
14. B. N. Green, J. V. Merren and J. G. Murray, American Society for Mass Spectrometry, 13th Annual Conference on Mass Spectrometry and Allied Topics, St Louis, Missouri (1965). Proceedings, p. 204.
Institute of Petroleum, London, 1967, p. 239.
15. W. J. McMurray, S. R. Lipsky and B. N. Green, in *Advances in Mass Spectrometry*, Vol. 4 (E. Kendrick, editor), Institute of Petroleum, London, 1967, p. 77.

CHAPTER 4

TREATMENT OF CALIBRATION DATA

For the reasons outlined in Chapter 3, such as spillage of sample, poor gravimetric and volumetric technique, poor injection technique, a leaking gas chromatograph septum, and poor reproducibility of mass spectrometric conditions, it is mandatory to use an internal standard to obtain good quantitative results. The use of an internal standard means that calibration curves must be constructed to cover the range of samples likely to be encountered for quantitative determination. At the present time, the most widely used internal standards of the types A, B and C defined in Chapter 3, are stable isotope labelled analogues of the compound to be determined. The construction of good calibration curves using these compounds requires attention to a number of points, many of which are relevant to calibration data for unlabelled internal standards of types B and C.

ISOTOPIC DISTRIBUTION IN A LABELLED SAMPLE

Although it is not strictly necessary to determine the concentration of the labelled isotope present in the internal standard before the latter can be used, a knowledge of the isotopic purity of the sample is important in deciding whether or not it will be a useful internal standard. It is shown later that a low isotopic incorporation can lead to trouble with calibration data. The number and nature of the isotopes present in the molecule must also be determined, and unless this is approached logically, difficulty can arise with multiply labelled compounds, such as those containing ^2H and ^{13}C simultaneously.

It is useful to compare the mass spectra of methane and two deuterium labelled methanes as representing a fairly simple case showing the effect of replacing hydrogen by deuterium on the mass spectrum. The spectra are given in Table 4.1.

As mentioned in the table heading, the spectra have been corrected for naturally occurring ^{13}C, that is, the contribution to each m/e value due to ions containing ^{13}C has been subtracted. The masses and natural abundance of important stable isotopes are given in Table 4.2.

TABLE 4.1
Mass spectra of methane, dideuteromethane and tetradeutero-methane. Spectra are corrected for ^{13}C contributions

	CH_4		$CH_2{}^2H_2$		C^2H_4	
m/e	Ion	$\% \Sigma_1$	Ion	$\% \Sigma_1$	Ion	$\% \Sigma_2$
20					C^2H_4	48.00
19						
18			$CH_2{}^2H_2$	45.38	C^2H_3	39.84
17			CH^2H_2	28.32		
16	CH_4	45.30	$CH_2{}^2H, C^2H_2$	13.93	C^2H_2	6.00
15	CH_3	39.00	CH_3, CH^2H	4.44		
14	CH_2	7.38	CH_2, C^2H	2.90	C^2H	3.53
13	CH	3.72	CH	1.27		
12	C	0.15	C	1.08	C	1.05
4			$^2H_2$	0.02	$^2H_2$	0.10
3			H^2H	0.08		
2	H_2	0.15	$H_2, {}^2H$	0.56	^2H	1.40
1	H	3.19	H	2.00		

The values given in the table for natural abundances are average values, since variations in nature occur according to the source of the particular element. Thus, the deuterium content of the oceans varies between 152 and 156 ppm, while rainwater can have a value as low as 133 ppm. [1]

TABLE 4.2
Masses and natural abundance of important stable isotopes

Element	Mass	Average abundance in atom $\%$	Element	Mass	Average abundance in atom $\%$
^1H	1.00783	99.985	^{29}Si	28.97649	4.70
^2H	2.01410	0.015	^{30}Si	29.97376	3.09
^{12}C	12.0	98.89	^{32}S	31.97207	95.05
^{13}C	13.00335	1.11	^{33}S	32.97146	0.76
^{14}N	14.00307	99.63	^{34}S	33.96786	4.22
^{15}N	15.00011	0.37	^{36}S	35.96709	0.014
^{16}O	15.99491	99.759	^{35}Cl	34.96885	75.53
^{17}O	16.99914	0.037	^{37}Cl	36.96590	24.47
^{18}O	17.99916	0.204	^{79}Br	78.9183	50.54
^{28}Si	27.97693	92.21	^{81}Br	80.9163	49.46

Because of the existence of natural abundance isotopes, the molecular ion region (and fragment ions) exists as a cluster. For a compound of composition $C_wH_xN_yO_z$ the expression

$$\left(\frac{100-c}{100} + \frac{c}{100}\right)^w \left(\frac{100-h}{100} + \frac{h}{100}\right)^x \left(\frac{100-n}{100} + \frac{n}{100}\right)^y$$

$$\times \left(\frac{100 - o_1 - o_2}{100} + \frac{o_1}{100} + \frac{o_2}{100}\right)^z$$

where c, h, n, o_1 and o_2 are the natural abundances given above for ^{13}C, ^2H, ^{15}N, ^{17}O and ^{18}O, gives the intensity ratio of the ions in the cluster. The values are obtained more conveniently from the tables of Beynon and Williams.[2] These values can be used to correct the intensities of a group of ions for natural abundance contributions.

The principle is as follows. Suppose we have a group of peaks in the spectrum of a hydrocarbon, as shown in Fig. 4.1. The ions occur at m/e 178, 179, 180 and

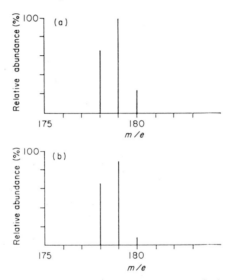

Fig. 4.1.   The region from m/e 175 to 184 of the spectrum of a hydrocarbon: (a) before correction; (b) after correction for the natural abundance of ^{13}C and ^2H.

181. Since there is no peak at m/e 177, the ion at m/e 178 is due entirely to $[C_{14}H_{10}]^+$ (based upon the known structure of the hydrocarbon) and contains no ^{13}C component. The intensities before correction are 65.2, 98.6, 23.2 and 2.5 respectively.

The tables of Beynon and Williams show that the ion of m/e 178 and due to $[C_{14}H_{10}]^+$ contributes 15.288% (mainly due to $[C_{13}{}^{13}CH_{10}]^+$) to m/e 179. The actual contribution is thus $15.288 \times 65.2 = 9.97$.

Therefore, the intensity at m/e 179 corrected for natural abundance is $98.6 - 9.97 = 88.63$ and is due to the ion $[C_{14}H_{11}]^+$.

This ion has a contribution to m/e 180 of 15.304% of its intensity, due mainly to $[C_{13}{}^{13}CH_{11}]^+$, the actual contribution being $15.304 \times 88.63 = 13.56$.

The ion at m/e 178 also makes a small contribution to m/e 180 due to the component $[C_{12}{}^{13}C_2H_{10}]^+$. From the Beynon and Williams' tables, this is equal to 1.09%, and therefore the contribution is $1.09\% \times 65.2 = 0.71$.

Thus, the corrected intensity at m/e 180 $= 23.2 - 13.56 - 0.71 = 8.93$.

The ion at m/e 179 has a contribution of 1.09% of its intensity to m/e 181. Therefore this is equal to 1.09% of 88.63 = 0.96.

The ion at m/e 180 has a contribution of 15.32% of its intensity to m/e 181. This is equal to 15.32% of 8.93 = 1.36%. These two components added together give an intensity of 2.32 for m/e 181, which compares with the value of 2.5 actually observed, so there is a very small component at m/e 181 of 0.18, due to $[C_{14}H_{13}]^+$.

Following these calculations, the corrected intensities of the ions in the region m/e 179 to 181 are 65.2, 88.63, 8.93 and 0.18 respectively.

This procedure of correcting for natural abundance contributions must be carried out before any calculation of isotopic incorporation can be undertaken.

Returning to the spectra in Table 4.1, a number of points can be made: the number of deuterium atoms in the molecule can be deduced simply from the increase in mass of the molecular ion; the partially deuteriated compound gives a more complex spectrum than the other two compounds; a mixture of methane and $[^2H_4]$methane would give a mass spectrum different from that of $[^2H_2]$-methane; fragment ions from each of the labelled compounds occur at the same m/e value as the molecular ion of unlabelled methane.

This last fact makes it extremely difficult to calculate the composition of a mixture of unlabelled and partially deuteriated compounds if the pure labelled compounds are not available. It would not be correct to take the highest m/e value (after correction of the whole group of peaks in the molecular ion region for natural abundance contributions as above), which represents the most highly deuteriated species, and calculate the contributions to lower masses due to fragmentation using the unlabelled compound as a model. This is because the unlabelled compound is not a model, since the probability of loss of hydrogen by fragmentation is not the same as the probability of the loss of deuterium. It has been shown frequently that the loss of hydrogen is greater than the loss of deuterium from partially deuteriated compounds.

If a pure sample of the various deuteriated components is available, the problem can be solved simply by a straightforward analysis of a multicomponent mixture as outlined in Chapter 5. In the absence of these compounds, the only way around the problem is to run the spectrum at a low enough electron beam voltage to prevent fragmentation by loss of hydrogen or deuterium, so that the molecular ion region consists solely of the molecular ions from the various labelled species. The ionization potential of isotopically labelled compounds is essentially the same as that for the unlabelled compounds, and so it can be concluded that the ratio of the intensities of the various ions represents the actual ratio of these components in the mixture.

Mixtures of deuteriated compounds are usually encountered when the labelled compound is made by an exchange reaction. This is exemplified by the two partial spectra given in Fig. 4.2(a) showing the molecular ion region of 2,6-dimethoxy-4-methylacetophenone before and after exchange in MeO^2H.

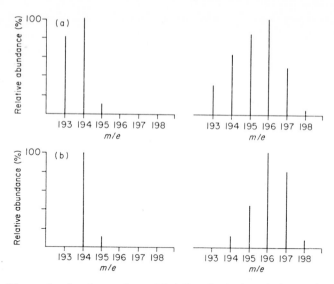

Fig. 4.2.   The molecular ion region of 2,6-dimethoxy-4-methylacetophenone before and after exchange in O-deuteromethanol: (a) at 70 eV; (b) at 11 eV.

The simplification achieved in the molecular ion region by running the spectra at low electron volts is shown in Fig. 4.2(b) where the 11-eV spectra are given. This enables the composition of the products ranging from $^2H_0$ to $^2H_3$ to be determined after correction for natural abundance.

Several assumptions have been made in outlining this method of determining the isotopic composition of the mixture: (a) the intensities per mole of the molecular ions of all the labelled species are the same; (b) there are no ion–molecule interactions giving rise to $[M + 1]^+$ peaks; if there were, these might vary from one species to another; (c) there is no fractionation or loss of labelled molecules in the mass spectrometer; (d) there is no contribution from background at the m/e values being measured; (e) the natural abundance peaks have the same intensity in the labelled compounds as in the unlabelled compounds. These may lead to some error in the determination of the isotopic distribution in certain cases. The first assumption may be violated in many examples for the following reasons.

When the abundance of the molecular ions of labelled and unlabelled species is measured, it is actually the difference between the number of ions of each species formed and the number of ions which subsequently decompose by fragmentation before passage through the magnetic sector or mass filter of the instrument that is being measured. Within the limits of experimental error it can be considered that the ionization efficiencies of labelled and unlabelled analogues are the same, thus producing the same numbers of ions from the same molar concentration of each compound. It is in the subsequent fragmentation stage

that differences arise, since primary and secondary isotope effects will be opera-
tional. The magnitude of these effects differs from one compound to another,
and so these cannot be compensated for by calculations based on experience
with other compounds. Their existence means that the relative intensities of the
molecular ions of labelled and unlabelled compound will not be identical for
the same amounts introduced into the mass spectrometer.

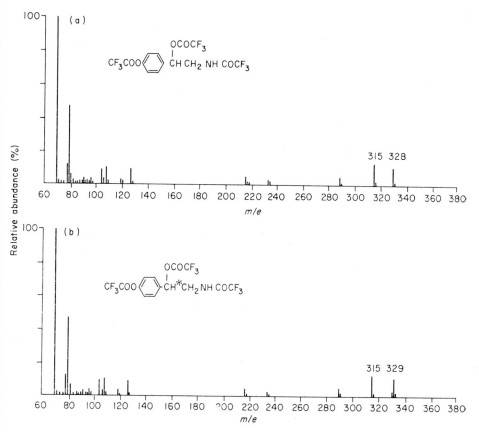

Fig. 4.3.   The mass spectrum of the trifluoroacetyl derivatives of: (a) octopamine;
(b) [^{13}C] labelled octopamine.

The primary isotope effects are differences in the cleavage of a bond between
a labelled atom and some other atom compared with cleavage of the corresponding bond in the unlabelled compound. A vast amount of data exists on the
primary deuterium isotope effect. As one example, ethanol has been shown
to lose a hydrogen atom specifically from the α-position. The spectrum of
α − [^{2}H]ethanol shows a ratio of intensities of $[M − H]^+/[M − {}^2H]^+$ of
3.57:1.[3]

In addition to the primary deuterium isotope effect, there may be the further complication of scrambling of the deuterium atoms over the molecule, frequently accompanied by the scrambling of carbon atoms. Thus, the $[C_7H_7]^+$ ion from toluene is completely scrambled prior to acetylene loss, as shown by studies with deuterium labelling and double ^{13}C labelling.[4,5] A general feature of scrambling reactions is that they become more extensive at low electron beam energies.

The error introduced by the primary isotope effect on the calculation of isotopic enrichment can be demonstrated for a $[^{13}C_1]$ labelled octopamine. The spectra of the unlabelled and $[^{13}C]$ labelled trifluoroacetyl derivatives (2 and 3) are shown in Fig. 4.3.

The manufacturers of the labelled material give an enrichment factor of 90.17 atom % for this compound.

The enrichment factor (e.f.) can be defined as follows. If I_L and I_U are the intensities of the molecular ions of the labelled and unlabelled compound respectively, then

$$\text{e.f.} = \frac{I_L}{I_U + I_L} \times 100 \text{ atom } \%$$

The molecular ions in the spectrum (obtained via a g.c.m.s. system) are so small as to introduce an unacceptable error into the calculation. The most convenient fragment ions to determine the enrichment factor are those having an m/e value of 329 in 3 and 328 in 2. The intensities of the ions m/e 327 to 329 are given in Table 4.3. In order to calculate an enrichment factor from the 70-eV

TABLE 4.3
Intensities of some fragment ions in the spectra of the trifluoroacetyl derivatives of octopamine and [^{13}C]octopamine

m/e	327	328	329	330
Octopamine trifluoroacetyl	8.0	101.0	12.0	1.1
[^{13}C]Octopamine trifluoroacetyl	0.1	22.0	98.0	10.0

spectra an assumption must be made about the m/e 327 ion. The assumption, which is not unreasonable, is that the extra hydrogen atom lost in order to form it is not subject to a secondary isotope effect, so that the labelled compound loses a hydrogen atom to the same extent as the unlabelled compound. In the former this will yield an ion of m/e 328, while in the latter there is the ion m/e 327. Thus, the intensity of m/e 328 in 3 is composed of three components: the natural abundance contribution due to ^{13}C from the m/e 327 ion (derived from the small amount of 2 present in 3), the loss of the extra hydrogen atom, either from the molecular ion or from m/e 329, and the unlabelled material 2 present in 3. It is the intensity of the latter which is required.

2 **3**

The contribution due to the loss of hydrogen from m/e 329 in the spectrum of **3** will be $\frac{8}{101} \times 98 = 7.76$ at m/e 328.

Since m/e 327 contains 12 carbon atoms, its contribution to m/e 328 due to the natural abundance of ^{13}C will be 0.013, which can be rounded off to 0.01. Hence, the intensity of m/e 328 due entirely to unlabelled **2** will be equal to $22.0 - 7.76 - 0.01 = 14.23$.

Therefore, the enrichment factor is

$$\frac{98}{14.23 + 98} \times 100 = 87.3\%$$

This figure is nearly 3% lower than that given by the manufacturers. The difference can be explained by the operation of a ^{13}C primary isotope effect, which will make the $\alpha-\beta$ bond less easily broken than in the unlabelled compound. Thus, the intensities of the ions due to the unlabelled material present in **3** will be greater than would otherwise be the case, making the compound appear to be less enriched.

The secondary isotope effect is the effect on the cleavage of a bond other than that between the labelled atom and some other atom. For example, the loss of a C^2H_3 group has been found to be less favoured than the loss of a CH_3 group from 1-[$^2H_3$]pentane, in which scrambling has been found to be almost negligible.[6]

Therefore, the total effect of the primary and secondary isotope effects is to decrease the intensity of fragment ions containing a heavy isotope compared with the unlabelled ions. As shown for [^{13}C]octopamine, calculations for the enrichment factors using such fragment ions can lead to values which are lower than the true value. On the other hand, for the reasons discussed earlier, the preferential fragmentation of unlabelled molecules will make the unlabelled molecular ion less intense than the heavy isotope containing molecular ions, so that calculations of enrichment factors for molecular ions will lead to values greater than the true value. *In view of this, the only safe course in calculating enrichment factors is to use only molecular ions, and to carry out the determination at electron voltages sufficiently low to prevent fragmentation.* In those cases where the intensity of the molecular ion is negligible, such as the octopamine example, one has either to accept that there will be an error, or the determination can be carried out under soft ionization conditions. At present, there appears to be no data in the literature concerning the determination of isotopic incorporation using chemical ionization.

The second assumption about $[M + 1]^+$ ions can lead to problems with some compounds. The main problem is the irreproducibility of the intensity of this ion from one run to the next, since its intensity depends on the square of the sample pressure, while the molecular ion intensity is directly proportional to the sample pressure. A long residence time in the source will increase the chance of an ion–molecule collision, leading to an $[M + 1]^+$ ion, so it is an advantage to operate at a high repeller potential to reduce the residence time. Care should be taken to run the labelled and unlabelled samples at the same sample pressure and repeller potential, and preferably as close together in time as possible.

As discussed in Chapter 5, where reservoir systems are used for the introduction of the sample, fractionation of the sample occurs in passing through the leak into the mass spectrometer source. The lighter molecules will pass through the leak preferentially, so that the sample in the reservoir contains an increasing proportion of the heavy isotope labelled molecules. Corrections for this behaviour are possible to a certain extent and are outlined in Chapter 5.

Exchange of isotope can occur in both the reservoir and the mass spectrometer source between the compound and water which is always present. This will happen to a considerable extent with groups such as O^2H, N^2H_2 and $CO_2{}^2H$. It is possible to equilibrate the instrument with ^2H_2O or CH_3O^2H before the determination, but the method is not reproducible. To all intents and purposes, compounds with such exchangeable deuterium atoms are of no use as internal standards for the measurement of the corresponding unlabelled compounds.

Compounds which contain deuterium atoms in apparently perfectly safe positions in the molecule can occasionally cause a surprise. The tetradeuteriated analogue (4) of prostaglandin $F_{2\alpha}$ (5) has been used widely for several years as a carrier and internal standard for the measurement of low levels of 5. The compounds are determined as the methyl ester tri-TMS derivative by monitoring the fragment ions at m/e 423 from the derivative of 5 and m/e 427 from the derivative of 4.

4   **5**

In these laboratories, a stock solution of **4** in methanol which had been used over a six-month period suddenly started to give high blank readings, i.e. for the usually low (<0.1%) amount of unlabelled material present as evidenced

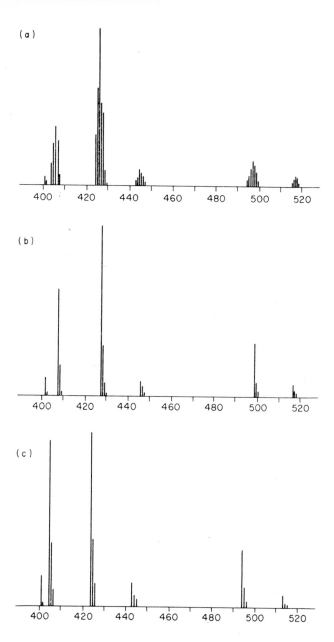

Fig. 4.4.   (a) Part of the mass spectrum of the tris-TMS derivative of 3,3,4,4-[^{2}H$_4$]-prostaglandin F$_{2\alpha}$ methyl ester after storage of the ester in methanol for some months prior to silylation. (b) The spectrum of the same derivative from a freshly prepared sample of the methyl ester. (c) The spectrum of the unlabelled analogue from freshly prepared methyl ester.

by the response at m/e 423. It was at first thought that some unlabelled prostaglandin $F_{2\alpha}$ had found its way into the stock solution inadvertently, but a determination of the mass spectrum of the methyl ester tri-TMS derivative (Fig. 4.4(a)) showed that this was not the case. The mass spectra of the same derivatives of 4 and 5 are shown in Fig. 4.4(b and c), and it can be seen that a simple mixture of the two would give responses only at m/e 423 and 427 in that region of the spectrum. The spectrum shown in Fig. 4.4(a) has responses at m/e 423, 424, 425, 426 and 427, so that it represents a mixture of compounds containing from 0 to 4 deuterium atoms. This means that exchange of deuterium with the methanol solvent must have occurred.[7]

The same phenomenon has been observed again recently in these laboratories when determining prostaglandin $F_{2\alpha}$ for another laboratory in samples and standards prepared by them. In the latest case, a stock solution of 4 had been kept in the refrigerator for about a year, portions being used monthly for the analyses. The stock solution suddenly developed the approximate isotopic composition of the labelled prostaglandin in the previous case.

The reason for this behaviour is not apparent at present, but it serves to sound a warning that labelled samples, especially if kept in solution, should be checked frequently for isotopic composition before embarking on time-consuming analyses.

It has been shown that it is possible to exchange enolizable hydrogen atoms with deuterium on g.c. columns containing Carbowax and KOD.[8] Therefore, it is possible that enolizable deuterium atoms may undergo exchange on some columns. It is always worth checking that the amount of label incorporated into the molecule remains the same by comparing a direct inlet spectrum with a spectrum obtained through the g.c.m.s. system.

The enrichment factor (e.f.) has already been defined in terms of the intensities of the respective molecular ions. It can also be defined in terms of the isotope ratio $R (= I_L/I_U)$ as follows:

$$\text{e.f.} = \frac{100R}{R + 1} \text{ atom } \%$$

The incorporation data can also be presented in the form of the isotopic abundance in excess of the natural abundance

$$\text{atom } \% \text{ excess} = 100 \left(\frac{1}{R_L + 1} - \frac{1}{R_U + 1} \right)$$

where R_L is the isotope ratio in the labelled compound and R_U is the isotope ratio in the unlabelled compound.

However, as pointed out by Campbell, there are two ways of calculating the atom $\%$ excess.[9] Suppose we have an enriched sample of CO_2 containing m ^{13}C atoms and n ^{12}C atoms. Let the natural abundance of ^{13}C in an unlabelled sample of CO_2 be p ^{13}C for every o ^{12}C atom, i.e. $100(p/p + o)\%$.

The definition given above for atom % excess would give

$$\text{atom } \% \text{ excess} = \frac{100m}{m+n} - \frac{100p}{p+o}$$

$$= 100\left(\frac{mo-pn}{(m+n)(p+o)}\right) = 100\left(\frac{mo-pn}{mp+mo+np+no}\right)$$

An alternative method is to remove from the m ^{13}C atoms in the enriched sample, the number (np/o) naturally associated with the n ^{12}C atoms, and then calculate the atom % excess as

$$\text{atom } \% \text{ excess} = 100 \; \frac{\begin{array}{c}\text{number of } ^{13}C \text{ atoms in sample,}\\ \text{excluding those occurring naturally}\end{array}}{\begin{array}{c}\text{total number of C atoms,}\\ \text{excluding those } ^{13}C \text{ occurring naturally}\end{array}}$$

$$= 100\left(\frac{m-np/o}{n+m-np/o}\right) = 100\left(\frac{mo-pn}{no+mo-np}\right)$$

The end results of these two methods are obviously different and values can be obtained to highlight the difference by using a simple example.

Take a sample of pure $^{13}CO_2$, where the enrichment factor is therefore 100 ^{13}C atom %. The atom % excess value will also be 100% by the second method, since no ^{12}C is present at that site to give rise to a natural abundance. The classical calculation gives by definition atom % excess = 100 − natural ^{13}C abundance = 98.931%. It is a matter of personal preference as to which of these alternatives is acceptable.

The simple computation involved in the application of Campbell's method can be illustrated by the determination of the enrichment factor for a sample of $[^{13}C_1]$ labelled phenacyl acetate $(C_{10}H_{10}O_3)$. The molecular ion region exhibited peaks at m/e 178, 179 and 180 with intensities in the ratio of 1:1.75:0.13. In a sample of unlabelled material, the corresponding ratios were 1:0.11:0.01. By subtraction the corrected ratios can be obtained:

$$m/e \; 178 \;\; 179 \;\; 180$$
$$1.00:1.75:0.13$$
$$0.11:0\text{'}01$$

$$\overline{}$$

$$1.00:1.64:0.12$$

In the equation e.f. = $100R/(R+1)$, R is therefore equal to 1.64, and so e.f. is $(100 \times 1.64)/2.64 = 62.1$ ^{13}C atom % excess.

Multiple incorporation

The preceding examples of ^{13}C labelled compounds have involved the incorporation of only one atom of the heavy isotope. However, in the case of deuterium labelled compounds there are usually several atoms incorporated into the molecule. This is partly because starting materials for the synthesis are cheaper when they are totally labelled (C^2H_3I, for example) and partly because it is necessary to keep to a minimum the amount of unlabelled material present when using these compounds as internal standards. This amount decreases as more heavy atoms are introduced into the molecule, and it can be calculated if the enrichment factor is known for each individual deuterium atom.

The simplest situation, which fortunately is the most frequently encountered, is where all of the deuterium (or any other heavy isotope) atoms have the same enrichment factor, say m atom %. If we incorporate n such atoms into the molecule, the molecular ion region consists of $(n + 1)$ peaks. Their relative intensity can be obtained from an expansion of the expression

$$\left(\frac{100 - m}{100} + \frac{m}{100}\right)^n$$

This will be familiar as the method of calculating the intensities of ions containing chlorine and bromine given in introductory textbooks to mass spectrometry.

If we introduce deuterium atoms with an enrichment factor of 99%, the expression becomes

$$(0.01 + 0.99)^n$$

For 1, 2 and 3 atoms introduced, the intensities of the molecular ion region are as follows:

	$[M]^{\ddag}$	$[M + 1]^{\ddag}$	$[M + 2]^{\ddag}$	$[M + 3]^{\ddag}$
1 deuterium	0.01	0.99		
2 deuteriums	0.001	0.0196	0.9801	
3 deuteriums	0.00001	0.0001186	0.029301	0.9703

As can be seen, the intensity of the molecular ion due to unlabelled compound rapidly becomes vanishingly small, being about 0·001% of the intensity of $[M + 3]^{\ddag}$ when three deuterium atoms are incorporated.

If numbers of isotopes are incorporated with different enrichment factors, a computation of the intensities in the molecular ion region involves an expansion of the expression

$$\left(\frac{100 - m_1}{100} + \frac{m_1}{100}\right)^{n_1}\left(\frac{100 - m_2}{100} + \frac{m_2}{100}\right)^{n_2}\left(\frac{100 - m_i}{100} + \frac{m_i}{100}\right)^{n_i} \times \cdots$$

where m_i is the enrichment factor of a particular isotope and n_i is the number of atoms of that isotope with that enrichment factor.

Obviously, the situation becomes extremely complex with a heavily labelled compound, and it may be difficult to obtain accurate enrichment factors by calculating back from the pattern of intensities in the molecular ion region, although a reasonable approximation can be made. It might be possible to choose fragment ions which do not involve scrambling of the labelled atoms but which reflect the isotopic distribution at particular points of the molecule. Although the calculations may then be subject to the errors outlined previously, the result may be statistically more significant than determinations based on molecular ion intensities. Alternatively, chemical degradation can be employed to reduce the compound to a number of products containing smaller numbers of labelled atoms whose incorporation can be determined more easily.

CALIBRATION CURVES

In any analytical technique, the value of one quantity y is usually measured for given values of another quantity x. There is normally very little error in x, but there may be a large error in y. In the field of mass spectrometry, for example, x could be a concentration per ml of a solution which has been extracted to provide a sample for determination, and y could be the response at a particular m/e value. Due to the error in the latter, a plot of y against x will show scatter about a curve or straight line. If it is assumed that the relationship should be linear, the straight line could be drawn by guesswork. However, a more accurate line can be constructed by the method of least squares. The result, called the regression line y/x, is the straight line which minimizes the sum of the squares of the vertical deviations from that straight line (Fig. 4.5).

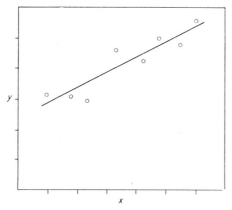

Fig. 4.5.   Regression line y/x.

The equation for the regression line is

$$y - \bar{y} = b(x - \bar{x})$$

where the slope b (called the regression coefficient) is given by

$$b = \frac{\Sigma\,(xy) - \Sigma\,(x)\,\Sigma\,(y)/n}{\Sigma\,(x^2) - \Sigma^2\,(x)/n}$$

and x, y are the means of a set of values (x_1, y_1), (x_2, y_2), etc. (see Appendix).

For the regression line of y/x, the assumption is made that the variance in y is much greater than the variance in x. If the variance in x is greater than that in y, a regression of x upon y must be carried out (see example at the end of this chapter, p. 86).

In general, before the above calculations are carried out, one must determine that there is indeed a significant correlation between the two sets of results x and y. This can be done by calculating the correlation coefficient r, given by

$$r = \frac{\Sigma (xy) - \Sigma (x) \Sigma (y)/n}{\sqrt{[\{\Sigma (x^2) - \Sigma^2 (x)/n\}\{\Sigma (y^2) - \Sigma^2 (y)/n\}]}}$$

If there is zero correlation between x and y, i.e. they are completely independent, $r = 0$. For perfect correlation between the two, $r = 1$ (see Appendix).

Statistical tables give values for r for different degrees of freedom ($= n - 2$) and probabilities. The calculated value of r must exceed the value from tables if there is to be a significant correlation. Thus, for five degrees of freedom (seven sets of x, y) at a probability level of 0·95, the theoretical value of r is 0.75. A calculated value of less than this means that there is no correlation between x and y.

The regression line will be calculated from values obtained using solutions of known concentration or samples of known weight. This then serves as a calibration line for determining the concentration or weight of the unknown. It is desirable to calculate the limits of error of the concentration or weight of the unknown so obtained from this calibration line. The appropriate equation is

$$V_x = \frac{V_y}{b^2} \left(\frac{1}{m} + \frac{1}{n} + \frac{(y - \bar{y})^2}{b^2} \times \frac{1}{\Sigma (x - \bar{x})^2} \right)$$

where V_x and V_y are the variances of x and y respectively, m is the number of repeat determinations, y is the value of the unknown and the other terms have the same meaning as in the regression equation. The limits of error can then be calculated from $\pm t \sqrt{V_x}$. The degrees of freedom used to obtain the value of t from statistical tables is $n - 2$, where n is the number of points on which the regression line equation has been based.

Most workers construct a calibration curve by making up several known mixtures of the compound to be determined plus the internal standard, and measuring the ratio of the mass spectrometric responses. The latter determination is carried out several times to increase the precision, and the regression line of response ratios on, say, mole ratio is calculated. This line of approach assumes therefore that the variance in response ratios is greater than the variance in mole ratios, i.e. the variance in the preparation of the standard solution. If this is true, it does indeed increase the precision of the measurements if several determinations of the response ratio for the same solution are carried out. However, it is to be regretted that most workers do not carry out an analysis of variance, as shown in Chapter 3, in order to check the basic assumption that the variance in response ratio is greater than that in solution preparation. If they did, they would almost certainly find, as shown by the measurements on dimethylallobarbitone, that the assumption is incorrect, and that the variance in making up the solutions is almost inevitably much greater than that in the mass spectrometric determinations. In this situation, *it is incorrect to carry out a regression of response ratios on mole ratios, but the reverse, a regression of mole ratios on response ratios should be determined*. To obtain greater precision, it is useless to analyse the same solution many times; instead, a number of solutions of the same mole ratio should be made up and analysed.

In using linear regression, it is assumed that the relationship between mole ratios and response ratios is linear. This is true only if one compound makes no contribution to the response to the other compound. For internal standards of type A, this means that the unlabelled standard contains no labelled species, while the labelled standard contains no unlabelled species. This situation is rarely true unless a large number, e.g. four, labelled atoms are incorporated in the internal standard, so that departure from linearity occurs, the degree of departure depending on the amount of isotopic impurity present in each standard. For a type C standard used in a direct inlet, the ion which is monitored from the internal standard must not be present in the spectrum of the compound to be measured, while the different ion monitored in the spectrum of the compound must not be formed by the internal standard. There is no problem, however, if the compounds are completely resolved in the g.c.m.s. system. If a variety of such type C standards is tested, it is usually possible to find one which will produce a linear correlation. Type B standards can be used only in a chromatographic inlet system, since the same ion is monitored for both compound and internal standard. Provided a complete chromatographic separation is obtained, there will be no cross-contribution to affect the linearity of mole ratio versus response ratio. Again, it is necessary to take reasonable care in the selection of a standard so that complete chromatographic resolution is obtained.

Type A standards

Before applying any mathematical treatment to deal with non-linearity, the degree of non-linearity should be determined, since it may turn out to be small enough to be ignored. Obviously, the simplest way is to plot the data and see whether a straight line results. A better way is to plot the difference between the observed response ratio and the response ratio calculated from the regression line and known mole ratio. Any curvature will result in increasing differences.[9] Several options are available to deal with non-linearity. Under certain conditions plotting inverse ratios can extend the linear range. Thus, for labelled standards, if the ratio of labelled to unlabelled species is plotted, the linear portion of the curve will be the region where the unlabelled species predominates. Conversely, when the ratio of unlabelled to labelled is used, the linear region will be that where the labelled species predominates.[10] The effect can be seen in Fig. 4.6.

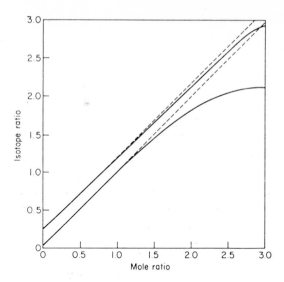

Fig. 4.6. Theoretical relationship for a mixture of natural abundance standard with 15% of the label isotope and a labelled standard with 2·5% unlabelled impurity. The lower curve is the plot of the ratio of the ion current observed for the labelled species to that of the unlabelled species, while the upper curve is the inverse. The dotted lines are extrapolations of the linear portions of the data. (Reproduced with permission from Ref. 10.)

Chapman and Bailey[11] have put forward the following treatment for non-linearity. Suppose the sample is monitored at m/e m_1 and standard at m/e m_2. Let the intensities of the contributions at m_1 and m_2 from sample be a_1 and a_2. Let the intensities of the contributions at m_1 and m_2 from an equal amount of internal standard be b_1 and b_2 respectively.

If the respective amounts of sample and standard in any run are A and B, then the measured ratio R is given by

$$R = \frac{a_1 A + b_1 B}{a_2 A + b_2 B}$$

If the contribution from the sample to mass m_2 is negligible, then $a_2 A \ll b_2 B$ and the relationship becomes

$$R = \left(\frac{a_1}{b_2 B}\right) A + \frac{b_1}{b_2}$$

which is of course linear.

However, if the standard itself is not isotopically pure, then $b_2 B$ will become smaller, depending on the degree of impurity. A state may be reached where $a_2 A$ is not negligible compared with $b_2 B$, so that R is no longer a linear function of A. The use of linear regression is then incorrect. However, it is possible to correct the relationship to linearity as follows:

R can be approximated by the quadratic equation

$$R = k_1 A^2 + k_2 A + k_3 \qquad (4.1)$$

A least squares fit may then be used to find the best values for k_1, k_2 and k_3 from the calibration values for R and A.

The actual calibration line used is the straight line fitted by the least squares method to a plot of A versus $R - k_1 A^2$.

From Eq. (4.1) A can be expressed by

$$A = \frac{B(b_2 R - b_1)}{(a_1 - a_2 R)}$$

so that in practice the calibration plot is

$$A \text{ versus } \left[R - k_1 \left(\frac{B(b_2 R - b_1)}{(a_1 - a_2 R)} \right)^2 \right]$$

Another approach for type A standards is to transform the mole ratio and isotope ratio data to yield a linear calibration.[9] This is done by calculating the atom % excess of each mixture and the labelled standard:

$$\text{atom \% excess} = 100 \frac{R}{R + 1}$$

where R is the isotope ratio which has been corrected for natural abundance contributions by subtracting the isotope ratio of the unlabelled standard from the isotope ratio of the diluted mixture.

Table 4.4 shows the relationship between the mole ratio, theoretical dilution, isotope ratio and atom % excess calculated for hypothetical mixtures of a

TABLE 4.4
Relationship between mole ratio, theoretical dilution, isotope ratio and atom% excess for a mixture of labelled and unlabelled standard, the unlabelled standard having a natural abundance contribution at the mass of the labelled standard of 15%. The labelled standard has 2·5% of the unlabelled material present[10]

Mole ratio of labelled/unlabelled	Theoretical dilution	Isotope ratio	Atom % excess
'Pure' labelled	1.0000	37.3322	97.3810
100.0	0.9901	27.0665	96.4179
30.0	0.9677	16.5205	94.2431
10.0	0.9091	7.8740	88.5373
3.0	0.7500	2.8615	73.0565
1.0	0.5000	1.1000	48.7180
0.333	0.2500	0.4722	24.366
0.100	0.09091	0.2472	8.8620
0.033	0.03226	0.1825	3.1448
0.010	0.00990	0.1598	0.9652
'Pure' unlabelled	0	0.1500	0

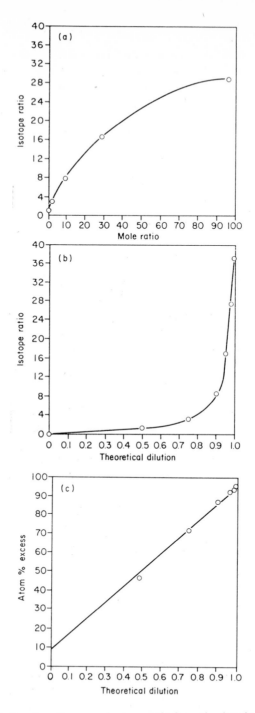

Fig. 4.7. (a) Plot of mole ratio versus isotope ratio from the data in Table 4.4. (b) Plot of theoretical dilution versus isotope ratio from the data in Table 4.4. (c) Plot of atom % excess versus theoretical dilution from the data in Table 4.4.

labelled standard which has a natural abundance contribution at the m/e value of the labelled standard and differing amounts of the labelled standard.[10]

The theoretical dilution is given by the ratio of moles of labelled standard to moles of labelled plus unlabelled standard. Figure 4.7(a) shows that a plot of mole ratio versus isotope ratio exhibits severe curvature, as does the plot of theoretical dilution versus isotope ratio shown in Fig. 4.7(b). In contrast, the plot of atom % excess versus theoretical dilution shown in Fig. 4.7(c) is linear.

A reliable calibration line should be obtainable by use of one of the approaches outlined above. The choice depends mainly upon the isotopic purity of the labelled and unlabelled standards.

A basic assumption made in carrying out linear regression is that the standard deviations of each population of y values are equal. According to Schoeller[10] this is the most frequently violated assumption in the application of regression statistics to isotope ratio calibrations. The standard deviation decreases as the isotope ratio decreases. Small isotope ratios have a smaller standard deviation than large isotope ratios (although they may have a larger *percentage* error). Therefore, they are defined with greater precision and should be given greater emphasis in the calculation of the regression line. This can be done by the use of weighting, which makes the regression less sensitive to the large absolute errors found at the upper extreme of the calibration. It also forces the regression line to pass close to the more precisely defined points near the origin.

The weighting factor used should be the inverse of the variance. Where the variance of the isotope ratio measurement is the major variance, the variance and so the weighting factors can be obtained from the calibration data for each solution. Where the major source of variance is sample preparation and dilution (this is more probable) the preparation variance must be obtained from the analysis of replicate solutions of the standards in order to obtain the weighting factor.

Although the comment has been made[11] that a practical limit of usefulness of isotopically labelled standards is when the contribution at the m/e value of the unlabelled compound is about twice that of the smallest sample to be encountered, the precision of the measurement must be taken into account. It is a significant difference between the means of two sets of results that is really the issue; those for the replicate isotope ratios of the pure labelled standard and those for the replicate isotope ratios of the labelled standard plus a small amount of added unlabelled material. A Student's t test is appropriate in this case, and the more precisely each set of results is defined, the closer the means can be while still retaining a significant difference in terms of t.

Because of this, a useful experiment to perform is to inject smaller and smaller amounts of the pure labelled standard, with perhaps four injections at each level. The coefficients of variation of the isotope ratio at each level can then be tabulated against the isotope ratio or amount of unlabelled compound represented by that level. Taking the case of the molecular ions $[M]^{+}$ and $[M + 3]^{+}$ from a trideuterated standard where the ratio of unlabelled material to labelled material is 1:50, a plot of coefficient of variation versus weight of unlabelled material injected gave the curve shown in Fig. 4.10.

As discussed earlier (p. 49), the variance in a ratio is more dependent upon the smaller component than on the larger component, so the general shape of the curve in Fig. 4.8 reflects the variance in the response to the small amount of

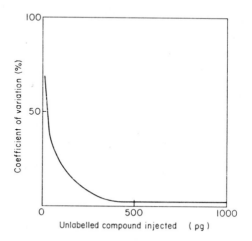

Fig. 4.8.   The injection of smaller and smaller amounts of a 50:1 mixture of labelled and unlabelled analogues. The coefficient of variation of the isotope ratio is plotted against amount of unlabelled material in the injection.

unlabelled material as a reasonable approximation. The curve enables us to estimate the coefficient of variation and hence the standard deviation involved in the determination of any amount of unlabelled compound which is present in the labelled standard from the detection limit upwards. The t test can then be used to determine the amount of extra unlabelled compound that has to be present at any point of interest along the curve to make the difference in the means significant.

The formula for t (see Appendix)

$$t = \frac{|m_A - m_B|}{s\sqrt{(1/n_A + 1/n_B)}}$$

can be rearranged to give

$$|m_A - m_B| \geq ts\sqrt{(1/n_A + 1/n_B)}$$

where m_A, m_B are the means of the sets of results A and B, n_A, n_B are the numbers of result in each set A and B, and s is the overall standard deviation.

The difference in the two mean results, one for the standard, and one for the standard plus the small extra amount of unlabelled compound, must exceed the expression on the right for the difference to be significant. Since four injections were used at each level, $n_A = n_B = 4$. The tabulated value for t ($P = 0.95$) is

2.45. Consider, for example, the point of the curve where an injection of 0.040 ng was used, the standard deviation s at this point being 0.00792. Putting this value with the other values into the equation shows that the difference between the two means must exceed 0.0137 for the difference to be significant. If the units of quantity are put in, this means that a significant difference in the response for 0.040 and 0.040 + 0.0137 (= 0.0537) ng injections can be observed. Thus, if the labelled standard contains 40 pg of unlabelled compound at the chosen injection level, it can still be used to detect 13.7 pg of unlabelled compound that will come from the extract to which the labelled standard is added.

The rearranged equation shows that the difference in the means which is significant is directly proportional to the standard deviation, so that in cases where the standard deviation is less than in the present experiment, an even smaller amount of additional unlabelled compound can be detected.

Evidence is now coming to light that many labelled compounds do not act as carriers for the unlabelled material through a g.c.m.s. system, and this is discussed in Chapter 6. In such a situation it makes no sense to use large ratios of labelled to unlabelled compound. If a reasonably high dynamic range of the compound of interest is to be covered, the amount of a type A labelled standard added to each unknown sample should be only a few times greater than the smallest level of unlabelled material likely to be encountered. This will lead to much lower limits of detection, since the contribution made by the small amount of unlabelled compound present in the labelled standard will be virtually negligible. Obviously, the calibration curve will be constructed using the same amount of standard and amounts of unlabelled compound sufficient to cover the whole range of concentrations required. For the higher concentrations of unlabelled compound, a response will be obtained from the unlabelled material which is much higher than the response from the much smaller level of internal standard present. This will introduce a rather larger variance for these higher concentrations than would be the case with larger amounts of internal standard, but the improvement at lower levels more than compensates.

The effect of using a large excess of a type A standard is shown in Fig. 4.9(a). The calibration line was constructed by injecting 10 ng of 3,3,4,4-[^{2}H$_4$]prostaglandin F$_{2\alpha}$ with varying amounts of the unlabelled compound (as methyl ester tris-TMS ether derivatives). Although the line is the result of linear regression, so that any possible curvature is ignored, it does not go through the origin because of the presence of a response at m/e 423 in the labelled compound, equivalent to about 0.4% of the response at m/e 427. Some of this extraneous response will be due to the presence of an appreciable amount of unlabelled material, while part could be due also to some other impurity which has an ion at m/e 423 and the same retention time. Thus, an injection of the labelled standard alone gives a response equivalent to the presence of about 40 pg of unlabelled material. If the precision of the measurement of the isotope ratio is poor, such a method cannot be used to measure much less than 20 pg or so of additional prostaglandin F$_{2\alpha}$.

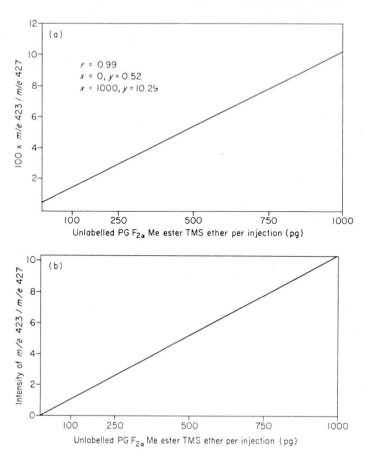

Fig. 4.9. (a) Calibration line for prostaglandin $F_{2\alpha}$ methyl ester tris-TMS derivative (PGF$_{2\alpha}$ Me TMS) using the type A internal standard 3,3,4,4-[^{2}H$_4$]PGF$_{2\alpha}$ Me TMS. Two-channel monitoring of the ions m/e 423 (from unlabelled compound) and 427 (from labelled compound) is used. The line is constructed from an injection of 10 ng of internal standard with varying amounts of prostaglandin from 1 ng down to zero. The line does not go through the origin because of the presence of about 0.4% of unlabelled PGF$_{2\alpha}$ as an impurity in the internal standard. It is assumed that the variance in the g.c.m.s. determination is greater than that in the preparation of standard solutions, so a regression of response ratio on amount of PGF$_{2\alpha}$ has been carried out. (b) Calibration line for prostaglandin $F_{2\alpha}$ methyl ester tris-TMS ether derivative, using the 3,3,4,4-[^{2}H$_4$] analogue as internal standard. The calibration curve is constructed with the same amounts of unlabelled prostaglandin as in Fig. 4.11, but with only 100 pg of the internal standard added. The line is now only an immeasurable amount from the origin. The regression of response ratios on amount of PGF$_{2\alpha}$ has been carried out as in Fig. 4.11.

The improvement which would be obtained by using a much lower ratio of internal standard is shown in Fig. 4.9(b). This theoretical calibration line is constructed by assuming that only 100 pg of the internal standard has been added to the same amounts of unlabelled prostaglandin as in Fig. 4.11. The unlabelled impurity present in this 100 pg amounts to 400 fg, so that essentially the line now goes through the origin. Moreover, provided that adequate sensitivity is available, it will be possible to quantify amounts down to a picogram or less.

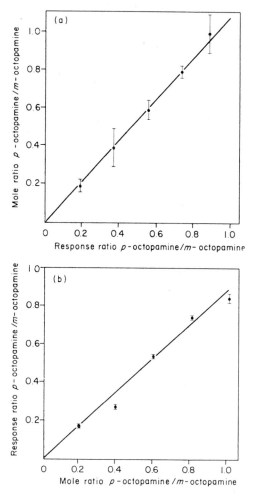

Fig. 4.10.   Calibration lines for the determination of p-octopamine as the tetrakis-TMS derivative, using m-octopamine as a type B internal standard, monitoring m/e 174.1. (a) The regression line for mole ratios on response ratios. The data were obtained from single determinations on each of three solutions prepared at each mole ratio. (b) The regression line for response ratios on mole ratios. The data were obtained from three determinations on single solutions at each mole ratio. Although the precision is poorer in (a) than (b), the accuracy is much greater for calibration line (a) than (b). This is because it was found that the error in the determination is much less than the error involved in preparing the solutions.

Type B standards

With a type B standard, which is a compound producing an ion in common with the compound of interest, the calibration line should pass through the origin, as shown in Fig. 4.10, unless there is any cross-contamination of the two components, or they are incompletely resolved on the gas chromatograph, or unless there is an impurity with a response at the same m/e value with the same retention time as the compound of interest.

Since it is obviously more economical to obtain a type B standard which is free of any of these problems than it is to obtain a type A standard which contains no unlabelled material (this requires the introduction of several labelled atoms), much better calibration lines should be obtained with the former type. The lines will go through the origin, and no correction for non-linearity will be necessary. In addition, since single ion monitoring can be employed, any mass spectrometer, however rudimentary, will be capable of this type of determination.

If a linear regression is used as described previously, the regression line may not pass through the origin but make a small intercept. Since the calibration line should pass through the origin, a different statistical approach must be used to ensure that this is the case.

If it is required to fit a straight line through the point (x_0, y_0) where there is a set of values (x_1, y_1), (x_2, y_2), etc. the slope of the regression line which minimizes the sums of squares of the vertical deviations is given by

$$b = \frac{\Sigma (x_i - x_0)(y_i - y_0)}{\Sigma (x_i - x_0)^2}$$

Thus, the slope b of the line which passes through the origin is given by

$$b = \frac{\Sigma x_i y_i}{\Sigma x_i^2}$$

This makes the assumption that the variance in y is greater than the variance in x (see Appendix).

The difference between the slope of the calibration line obtained by this method and a normal linear regression is illustrated for some response ratios as follows:

Picograms of compound injected	0	400	800	1200	1600	2000
Response ratio, compound/standard	0	0.19	0.37	0.56	0.75	0.91

The slope calculated by a normal linear regression is 0.00045 and the intercept 0.00476, while the slope calculated by the above method is 0.000462, and the intercept zero.

Change in slope of the calibration line

The slope in a calibration line almost certainly changes if the g.c. conditions are changed, for example the flow rate and temperature. A change in the type of column used can lead to a gross change in the calibration line. For a 1:1 mixture of *m*- and *p*-octopamine, for example, the response ratio of *para* to *meta* in the TMS derivatives was 0.94, this being the mean of three injections on a 1 % SE 30 column at 210 °C. When the column was changed to 2 % Dexsil, the response ratios on exactly the same mixture averaged 0.38, so that the slope of the calibration line decreased by almost a factor of three in changing columns. This must be the result of greater absorption or decomposition of the *para* derivative on Dexsil compared with SE 30, thus decreasing the response for *para*, while the *meta* derivative is not affected to the same extent.

A day to day variation may also be observed on the same column run at the same temperature and flow rate, so that the validity of the calibration line should be checked daily before embarking upon the determination of unknowns using the calibration. Two approaches can be used to test for a change in the calibration. First, it is often possible to construct the original calibration line from aliquots of bulk mixtures of compound and its internal standard (which may or may not require evaporation and derivatization). Thus, exactly the same solution can be analysed at the start of each day. A Student's *t* test can then be used to see whether the mean response ratio of a number of injections is significantly different from the mean response of the same number of injections used in the construction of the calibration line. A significant difference at the 95 % confidence level means that the old calibration line should be discarded.

The second approach is when the same solution cannot be preserved for re-evaluation each day. In this case, a new solution must be prepared and analysed, and the difference between the observed and predicted response ratio compared with the standard deviation at that point of the calibration line using the *t* test. The standard deviation of a point determined from the calibration line will depend upon which regression was used. If it was determined that the variance in the g.c.m.s. determination was greater than the variance in solution preparation, as discussed in Chapter 3, then the calibration line will be a regression of response ratios (y) on mole ratios (x). The standard deviation of a mole ratio s_{mr} determined from such a calibration line (see Appendix) is given by

$$s_{mr} = \frac{s_{rr}}{b} \left(\frac{1}{m} + \frac{1}{n} + \frac{(y - \bar{y})^2}{b^2 \Sigma (x - \bar{x})^2} \right)^{\frac{1}{2}}$$

where s_{rr} is the standard deviation of the response ratios, m is the number of determinations of the unknown, n is the number of points used for the regression, y is the value of the response ratio of the unknown and the other terms have the same meaning as in the equation for the regression line discussed earlier.

If it was shown that the variance in the solution preparation was greater than that in the g.c.m.s. determination, the calibration line will be a regression of mole ratios (y) on response ratios (x). The formula for the standard deviation of a mole ratio s_{mr} is then

$$s_{mr} = s_{rr} \left(1 + \frac{1}{n} + \frac{(x - \bar{x})^2}{\Sigma (x - \bar{x})^2} \right)^{\frac{1}{2}}$$

Construction of a calibration line for octopamine

As a practical example, the approach used in these laboratories for the construction of a calibration line for the measurement of p-octopamine using m-octopamine as the internal standard is outlined below.

Instead of adding m-octopamine to cerebrospinal fluid (CSF) samples and then extracting these with a solvent to determine the ratio of p-octopamine to the internal standard, which would be the more usual approach, the p-octopamine and the internal standard were isolated by liquid chromatography before lyophilizing and forming the TMS derivatives. It was determined by studies on radioactive octopamine that over 90% was consistently recovered in this way, so the calibration line could be constructed by making up known mixtures of p- and m-octopamine, forming the TMS derivatives and determining the responses for the common fragment ion at m/e 174.1. The two compounds are resolved on the g.c. column, and neither pure compound is contaminated with traces of the other.

The first step is to decide whether the largest source of error in the determination lies in the making up of the solutions or in the g.c.m.s. determination. Consequently three mixtures were made up of p- and m-octopamine containing 1.2 ng and 2 ng respectively of each compound as the TMS derivatives in a 1-μl injection on the gas chromatograph. Each solution was determined three times, with the results shown in Table 4.5.

TABLE 4.5
The response ratios for p-octopamine to m-octopamine for three solutions each containing nominally 1.2 ng of p- and 2 ng of m-octopamine

Solution	Response ratios			Mean ratio	Standard deviation
1	0.52	0.52	0.52	0.52	0
2	0.62	0.62	0.62	0.62	0
3	0.54	0.54	0.54	0.54	0

This is a case where it is hardly necessary to carry out an analysis of variance as outlined in Chapter 3. The variance in the g.c.m.s. determination is zero, while quite obviously there is a variance between the solutions, since the means are not identical for solutions which should contain the same amounts. In this situation, it is pointless to carry out replicate injections of the same solution, since they will all give the same result. Therefore, the correct approach was to make up three solutions for each particular mixture and carry out one determination on each. The results are shown in Table 4.6.

TABLE 4.6
Responses for three solutions made up at a number of mole ratios of *p*-octopamine to the internal standard *m*-octopamine

Mole ratio	Responses			Mean	Standard deviation
0	0	0	0	0	0
0.2	0.18	0.22	0.16	0.186	0.03
0.4	0.39	0.45	0.27	0.37	0.09
0.6	0.52	0.62	0.54	0.56	0.05
0.8	0.74	0.70	0.75	0.73	0.026
1.0	0.81	1.0	0.84	0.88	0.10

Since it has been shown that the solution variance is greater than the g.c.m.s. variance, a regression of mole ratios (y) on response ratios (x) must be carried out.

If the calibration line is not required to go through the origin, a normal calculation of the regression line gives the following:

$$\text{slope} = 1.09055 \qquad \text{intercept} = 3.8 \times 10^{-3}$$

It is interesting to compare this with the results obtained if the (incorrect) regression of response ratios on mole ratios is carried out:

$$\text{slope} = 0.89095 \qquad \text{intercept} = 9.52 \times 10^{-3}$$

Since the two different regressions have the axes interchanged, the reciprocal of the second slope must be taken in order to compare it with the first.

The reciprocal of 0.89095 is 1.122, so the slope of this line is not quite the same as the correct regression line of mole ratios on response ratios.

As stated earlier (p. 84), since there was no cross-contamination of the two compounds, *m*-octopamine on its own (mole ratio = 0 in Table 4.6) gave a zero response ratio. Therefore the calibration line should pass through the origin. This can be brought about by the statistical approach outlined previously (see also Appendix), which assumed that the variance in y was greater than the variance in x. Therefore y must be the mole ratios and x the response ratios.

The slope of the calibration line, b, is given by

$$b = \frac{\Sigma \, (\text{response ratio})_i (\text{mole ratio})_i}{\Sigma \, (\text{response ratio})_i{}^2}$$

$$= \frac{\Sigma \, (0.18 \times 0.2) + (0.22 \times 0.2) + (0.16 \times 0.2) + \cdots}{\Sigma \, 0.18^2 + 0.22^2 + 0.16^2 + \cdots + 0.81^2 + 1^2 + 0.84^2}$$

$$\phantom{= \frac{\Sigma}{\Sigma}} + (0.81 \times 1) + (1 \times 1) + (0.84 \times 1)$$

$$= 1.0963$$

The slope of this regression line which passes through the origin is only marginally different from that obtained by the usual linear regression, but other examples may be obtained in which the difference is much more marked.

If the calculation of the slope had been carried out incorrectly by making y the response ratios and x the mole ratios, the slope would have been

$$\frac{\Sigma \text{ (response ratios)}_i \text{(mole ratios)}_i}{\Sigma \text{ (mole ratios)}_i{}^2} = \frac{5.966}{6.6} = 0.9039$$

The reciprocal of this, 1.1062, is slightly different from the slope of the correct calibration line obtained previously.

The calibration line, with a slope of 1.096, passing through the origin and obtained by a regression of mole ratios on responses is shown in Fig. 4.13(a). In order to transpose the standard deviations, which are in the responses originally, into corresponding standard deviations of the mole ratios, they have been multiplied by the slope of the line.

If the experiment of preparing a number of replicate solutions of the same nominal mole ratios had not been carried out, the data would have consisted of triplicate responses on one solution at each mole ratio. The data from such an experiment are presented in Table 4.7.

TABLE 4.7
Responses for three determinations on one solution of each mole ratio of p-octopamine to the internal standard m-octopamine

Mole ratio	Responses			Mean	Standard deviation
0	0	0	0	0	0
0.2	0.16	0.18	0.17	0.17	0.01
0.4	0.27	0.27	0.27	0.27	0
0.6	0.54	0.53	0.54	0.54	0.01
0.8	0.75	0.75	0.76	0.75	0.01
1.0	0.84	0.84	0.83	0.84	0.01

A regression of responses on mole ratios can then be carried out. Since the line must go through the origin, the slope b can be calculated from

$$b = \frac{\Sigma \text{ (response ratios)}_i \text{(mole ratios)}_i}{\Sigma \text{ (mole ratios)}_i{}^2}$$

and comes out as 0.8642.

This calibration line is shown in Fig. 4.10(b) along with the standard deviation of the points. Several observations can be made about the two calibration lines in Fig. 4.10. Although the precision for each mole ratio in Fig. 4.10(a) is poor, the regression line passes extremely close to the means at all mole ratios, while in Fig. 4.10(b) the precision for the points for each mole ratio is very high, reflecting the high precision of the g.c.m.s. determination. However, the scatter of the means of these points about the regression line is very high, because of the error involved in making up the standard solutions. Any unknown solution interpolated on calibration line (a) will give a much more accurate value for the amount of compound present than an interpolation on calibration line (b).

As in the octopamine case, the correct calibration line for a situation in which there is a greater error in the preparation of solutions than in the g.c.m.s. determination will have response ratios along the x-axis, since the normal method of carrying out a regression assumes that x has the least variance. A study of the literature shows that the vast majority of calibration lines are presented with mole ratios, or ng per ml or some other measurement of quantity along the x-axis. Therefore, one would conclude that a regression of response ratios on mole ratios has been carried out. Since most mass spectrometers in good condition are capable of precisions of the order of $1-2\%$, at least for injections of a few nanograms, it is probable that in almost every case the error in the preparation of solutions is higher than this.

In addition to the solution preparation error, there is also the fact that much poorer precision is obtained when calibration lines are constructed by means of spiked solutions, as mentioned in Chapter 3. In the octopamine case, spiking was not adopted for two reasons. First, the liquid chromatography method was superior to extraction, since it gave reproducible recoveries, the samples so obtained being relatively free of other interfering substances. Second, it was not possible to obtain CSF free of p-octopamine, so that a simple spiking experiment would not suffice. It would have been necessary to adopt a method of standard additions of increasing amounts of p-octopamine (including a zero) plus a fixed amount of internal standard in order to determine the amount of p-octopamine originally present.

For those instances, mainly in drug determination, where a blank urine or plasma sample can be obtained which does not contain the compound of interest, calibration lines should be constructed from samples spiked with the compound and its internal standard. The error in such a determination will then almost certainly be higher than the g.c.m.s. determination alone. Thus, calibration lines produced by such a method in which only one extraction has been performed to determine a point on the calibration line by replicate injections have been determined incorrectly, although in some cases the difference between the regressions might be insignificant. Unfortunately, it is very rare for data to be presented to enable a judgement to be made of which is the greater source of error, but it is to be hoped that a more critical approach to the determination of calibration lines will evolve over the next few years.

REFERENCES

1. L. Friedman, *Geochim. Cosmochim. Acta* **4**, 89 (1953).
2. J. H. Beynon and A. E. Williams, *Mass and Abundance Tables for Use in Mass Spectrometry*, Elsevier, Amsterdam, 1963.

3. M. Corval, *Bull. Soc. Chim. Fr.* 2871 (1970).
4. I. Howe and F. W. McLafferty, *J. Am. Chem. Soc.* **93**, 99 (1971).
5. A. S. Siegel, *J. Am. Chem. Soc.* **92**, 5277 (1970).
6. B. J. Millard and D. F. Shaw, *J. Chem. Soc. B* 664 (1966).
7. T. H. Cory, P. T. Lascelles, B. J. Millard, W. Snedden and B. W. Wilson, *Biomed. Mass Spectrom.* **3**, 117 (1976).
8. M. Senn, W. J. Richter and A. L. Burlingame, *J. Am. Chem. Soc.* **87**, 680 (1965).
9. I. M. Campbell, *Bioorg. Chem.* **3**, 386 (1974).
10. D. A. Schoeller, *Biomed. Mass Spectrom.* **3**, 265 (1976).
11. J. R. Chapman and E. Bailey, *J. Chromatogr.* **89**, 215 (1974).

CHAPTER 5

QUANTIFICATION USING NON-CHROMATOGRAPHIC INLET SYSTEMS

Non-chromatographic inlet systems may be defined as hot and cold reservoir systems, where the sample passes into the mass spectrometer source via a leak, and direct inlet systems, in which the solid sample is introduced on a probe, that may or may not be heated independently of the source.

In order that an analysis of the mixture entering the mass spectrometer from such inlets can be carried out with the minimum of mathematical manipulation, several criteria should be satisfied.

(a) The mass spectrum of each component should be reproducible throughout the period of time required for the analysis of standard and unknown mixtures.

The influence of the various factors which can affect reproducibility has been covered in Chapter 3. A logical sequence should be followed for tuning the ion source to maximum reproducibility. Most magnetic instruments have a number of controls which are meant to be adjusted in order to tune the source. These include ion repeller, beam centring, and vertical and horizontal deflectors. One approach to tuning the source is to maximize the total ion current reading, while another is to maximize the intensity of a peak being received at the collector, the most useful peak being one in the middle of the mass range which needs to be scanned for the quantitative work. Occasionally these result in the same source settings, but more often, having tuned the total ion current to maximum, it will be necessary to alter one or more of the controls to maximize a peak at the collector. The result of these adjustments should be the production of symmetrical peaks which are reproducible in intensity from one scan to another, assuming a constant sample level.

An important piece of equipment for the mass spectroscopist is a storage oscilloscope, which can be used to check that peaks are symmetrical in shape, and also that both the shape and intensity are reproducible. This can be carried out by setting the trigger on the oscilloscope so that the first peak over a certain intensity in the mass spectral scan causes the oscilloscope to sweep. The scan

can be repeated several times so that the same peak is superimposed on its previous image. Any change is then immediately obvious.

The controls which affect peak shapes and intensities in quadrupole instruments are usually much less accessible. Tuning these is rather straightforward, however, because nearly all commercial quadrupole instruments have a built-in oscilloscope. Since a large number of quadrupole instruments being sold at present also have dedicated data systems, the computer is beginning to take over the job of routinely tuning up the instrument. This is done by carrying out several iterations of the tuning procedure with PFK or perfluorotributylamine in the source until the intensities of selected ions are within the specified limits.

(b) The mass spectrum of one component should be unaffected by the presence of any other component.

This means that the contributions of each compound to a given m/e value in the total mass spectrum should be linearly additive.

(c) The abundance of ions used for the analysis should be proportional to the partial pressure of that component.

An obvious situation when (b) and (c) may not be true is when the pressure in the source is unduly high. Above about 10^{-5} Torr ion–molecule reactions become important. In the case of organic compounds, the most common reaction is hydrogen abstraction by molecular ions from neutral molecules. For a mixture, therefore, important factors are the ease of abstraction of a hydrogen radical from each component, and the proportion of each component in the mixture.

The occurrence of ion–molecule collisions in an e.i. source may be recognized by running the spectra at a series of known sample pressures. Normal mass spectral fragmentation is unimolecular in nature, so that peak intensities are proportional to sample pressure. Ion–molecule reactions are bimolecular and lead to peaks whose intensities are proportional to the square of the pressure. Any peak whose intensity is seen to be proportional to the square of the pressure should not be used for quantitative work. A simpler method of detecting peaks formed partly by ion–molecule collisions is to change the repeller voltage. This directly affects the residence time of ions in the source, and therefore the probability of collisions between ions and neutral molecules. As a result, the intensities of the resultant peaks will change more than those of normal fragment ions. Again, such peaks should not be used for quantification.

SAMPLE ADMISSION WITHOUT FRACTIONATION

The majority of cases where hot and cold reservoir systems are used can be included under this heading. It can be assumed that the leak which is placed between reservoir and source allows very little differential passage of components into the source. Thus, the composition of the mixture entering the source is taken to be the same as that in the reservoir. Although this is never strictly true, for most cases this assumption leads to only a small error which may be neglected.

In order to carry out the quantitative analysis of a mixture which has been introduced via a reservoir system, it is necessary to run the spectra of the pure components. In the early days of such analyses, the pure components were run at known pressures, in order to determine the sensitivity coefficients. The sensitivity coefficient is the response at an m/e value of interest from a pure component at unit pressure.

The analysis then depends upon solving a series of simultaneous equations which involve sensitivity coefficients for each component:

$$S_{11}P_1 + S_{12}P_2 + \cdots + S_{1n}P_n = I_1$$
$$S_{21}P_1 + S_{22}P_2 + \cdots + S_{2n}P_n = I_2$$
$$\cdots \cdots \cdots \cdots \cdots \cdots \cdots$$
$$S_{m1}P_1 + S_{m2}P_2 + \cdots + S_{mn}P_n = I_n$$

I_m represents the intensity of a peak at an m/e value m in the spectrum of the mixture. P_n is the partial pressure of the component n, while S_{mn}, the sensitivity coefficient, is the intensity of the peak at m/e value m in the spectrum of the pure component n, i.e. n is at unit pressure.

Thus, there are m simultaneous equations in the n unknowns. When the number of equations equals the number of unknowns the usual methods can be applied to solve the equations for partial pressures P_n. A least squares treatment can be applied if the number of equations exceeds the number of unknowns.

The great disadvantage of the method as outlined above is that it requires a knowledge of the pressure of each of the pure components when their spectra are being run so that the sensitivity coefficients S_{mn} can be determined. It is not easy to measure such pressures accurately. In the simple case of a binary mixture, mole fractions can be used instead of pressures. Mole fractions are determined quite simply, and this only needs to be done once, when making up a known binary mixture.

Suppose the ratio of the partial pressures (derived from the known mole fractions) in the known binary mixture is P_1/P_2 and the ratio of partial pressures in the unknown binary mixture is P'_1/P'_2, then, using simultaneous equations as before, for the known mixture

$$S_{11}P_1 + S_{12}P_2 = I_1$$

$$S_{21}P_1 + S_{22}P_2 = I_2$$

thus

$$\frac{S_{22}}{S_{21}} = \frac{(I_1/I_2 - S_{11}/S_{22})}{(S_{12}/S_{22} - I_1/I_2)} \times \frac{P_1}{P_2} \tag{5.1}$$

and for the unknown mixture

$$S_{11}P'_1 + S_{12}P'_2 = I'_1$$

$$S_{21}P'_1 + S_{22}P'_2 = I'_2$$

so that

$$\frac{P'_1}{P'_2} = \frac{(S_{12}/S_{22} - I'_1/I'_2)}{(I'_1/I'_2 - S_{11}/S_{21})} \times \frac{S_{22}}{S_{21}} \tag{5.2}$$

substituting (5.1) into (5.2) gives

$$\frac{P'_1}{P'_2} = \frac{S_{12}/S_{22} - I'_1/I'_2}{I'_1/I'_2 - S_{11}/S_{21}} \times \frac{I_1/I_2 - S_{11}/S_{21}}{S_{12}/S_{22} - I_1/I_2} \times \frac{P_1}{P_2} \tag{5.3}$$

In this equation the ratio S_{11}/S_{21} represents the ratio of the intensities of the two relevant peaks in the spectra of the pure component 1 and S_{12}/S_{22} the ratio of these peaks in the spectrum of pure component 2.

The sequence of determinations needed to solve the composition of an unknown binary mixture is therefore: (a) run the mass spectra of pure components 1 and 2; (b) choose two ions and establish the ratio of their intensities in each of the two spectra; (c) make up a mixture of known mole fraction and determine the ratio of the same two ions in this mixture; (d) finally, run the spectrum of the unknown mixture and establish the ratio of the two ions in this.

This gives sufficient data to solve Eq. (5.3) for partial pressures or mole fractions of the components in the unknown binary mixture.

As can be imagined, the above equations become cumbersome when a multi-component mixture is being handled, since large determinants will be needed. However, the use of computer programs can reduce the tedium of such calculations.

A successful analysis depends upon the correct choice of ions from the mass spectrum. Thus, for a binary mixture, the ratio of the two ions monitored is most sensitive to a change in the ratio of the components if there is no cross-contribution, i.e. component 1 has a large response at the first mass and no response at the second mass, while the reverse is true for component 2. At the other extreme, zero sensitivity to a change in ratio of the two components is obtained if both have identical contributions at each mass.

The dependence of the sensitivity of the ratio to composition changes upon the cross-contribution can be illustrated as follows. Consider two pure components 1 and 2 and assume that the ratio of the intensities of the two ions chosen for the analysis is equal but opposite for the two components when the molar concentrations are identical. Although this is an unreal situation, it simplifies the calculation and enables a qualitative picture to emerge on the effect of varying the intensity ratios. In this situation, as shown in Fig. 5.1, a 1:1 mixture of the two components will have identical responses at the two masses chosen for the analysis. The change in these two responses from 50:50 in the mixture must then be determined when the mixture composition is changed to 55:45. Figure 5.1 shows how the spectrum of this new mixture appears for four different cases, where there is an infinite, 10:1, 2:1 and equal ratio of peak intensities in the original pure components.

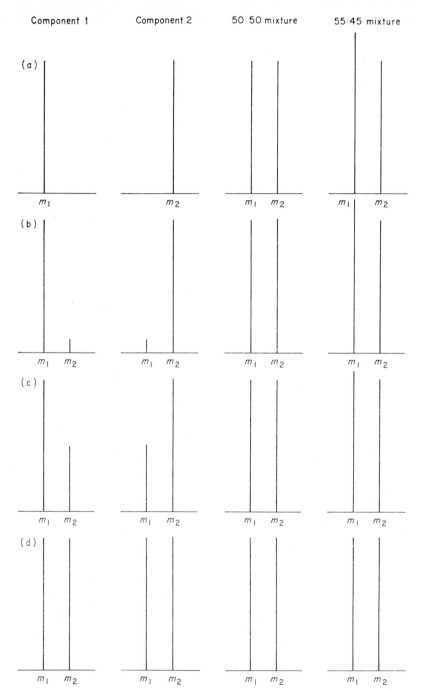

Fig. 5.1. Appearance of two ions in the spectra of component 1, component 2, and 50:50 and 55:45 mixtures of the both: (a) when the ratios of the ions are 0:1 and 1:0 in components 1 and 2 respectively; (b) when the ratios are 10:1 and 1:10; (c) when the ratios are 2:1 and 1:2; (d) when the ratios are 1:1 and 1:1.

The measured response ratios for these four cases plus three others are given in Table 5.1.

TABLE 5.1
Ratio of the response at two m/e values in 50:50 and 55:45 binary mixtures for various ratios of these responses in the pure components. The ratio of the responses in component 1 is equal and opposite to that in component 2

Response ratio at m_1/m_2 in component 1 (and m_2/m_1 in component 2)	Response ratio in 50:50 mixture	Response ratio in 55:45 mixture
∞	1.0000	1.222
20	1.0000	1.198
10	1.0000	1.178
5	1.0000	1.143
4	1.0000	1.128
2	1.0000	1.068
1	1.0000	1.000

The ∞ value is the case where there is no cross-contribution, so the response ratio of 1.222 is identical with the component ratio of 55:45 ($= 1.222$). For this value, a change in component ratio from 1.222:1 leads to exactly the same change in response ratio, 1.222:1. At the other extreme, where the response ratio for pure components is 1, the response ratios for both the 50:50 mixture and the 55:45 mixture are identical at 1. In such a situation, obviously the analysis fails since any mixture of the two components gives the same responses. The intermediate points show a smaller and smaller change in the response ratio for the mixture in changing from a 50:50 situation to a 55:45 situation as the response ratios in the pure components are decreased. Table 5.1 shows that for peak ratios of 4:1 and 1:4, the change in the response ratio is about half of the change in mole ratio—1.128 from 1.000 as opposed to 1.222 from 1.000. To achieve any kind of sensitivity in the response ratio to changes in the mole ratio of components, peak ratios of 2:1 probably represent the lower limit at which such an analysis is worthwhile.

Of course, not only the change in response ratios is important, but also the error with which the peak height can be measured. It is possible to show that the error in the peak height measurement is far greater than the error due to ion statistics if a stable sample level is assumed. For quantitative work using reservoir systems, a study of the literature shows that the mass spectrometer is used exclusively in the scanning mode. The scan speed and resolving power are therefore of crucial importance, since these determine the number of ions recorded on each peak of interest. The use of a reservoir system also implies that a fairly large amount of sample is available. Under these conditions, the maximum flow rate of sample into the source is probably about 100 ng s^{-1}.

If the sample is assumed to have a molecular weight of 200, this gives a potential number of ions as

$$\frac{10^{-7} \times 6.023 \times 10^{23}}{200}, \text{ approximately } 3 \times 10^{14} \text{ ions}$$

It can be assumed that the efficiency of the mass spectrometer source is about 0.1%, so the number of ions produced per second $= 3 \times 10^{11}$.

If it is further assumed that the peaks monitored each contain 10% of the total ion current, then the number of ions in each of these peaks per second will be 3×10^{10}.

The time spent per peak in an exponential scan is given by

$$t = \frac{0.43 t_{10}}{R_s}$$

where t_{10} is the time taken to scan a decade in mass and R_s is the static resolving power.

For a typical scan of 10 s per decade at a resolving power of 1000, the number of ions obtained in the peaks being monitored will be

$$\frac{0.43 \times 10}{1000} \times 3 \times 10^{10} = 1.29 \times 10^8$$

The coefficient of variation due to ion statistics $= \dfrac{100}{\sqrt{(\text{number of ions in peak})}}$

$$= \frac{100}{\sqrt{(1.29 \times 10^8)}} = 0.0088\%$$

For the purposes of the present discussion this is negligible when it is considered that a peak can only be measured on the chart spectrum to within about ±0.5 mm. The error due to measurement of a peak which is, say, 1 cm high on the chart is therefore much larger than that due to ion statistics.

The overall effect of this error in peak height measurement becomes rather complicated, since it must be taken into account in measuring up the spectrum of the pure components as well as that of the mixture. As shown earlier (p. 94), in order to make the analysis as sensitive as possible to composition changes, peaks are chosen purposely in each spectrum so that one is very intense and one is very weak. The measurement of the weak peaks will obviously be subject to a large error which will be carried through the mathematical analysis. If peaks which are not so different in intensities are used, the errors in measurement will be reduced, but the analysis will be less sensitive to changes in the mixture composition. Taking all these factors into account, the intensities of the peaks chosen for the analysis should lie somewhere between 2:1 and 10:1.

The points raised in the discussion of the analysis of binary mixtures are equally valid when applied to multicomponent mixtures. For the analysis of an *n*-component mixture, a minimum of *n* peaks will have to be chosen from the spectrum of each component. The principle of choosing a peak which is as unique as possible to that component will still apply. However, there comes a point where the time taken to choose these peaks and then measure them, in addition to the calculation, makes such an analysis uneconomical by mass spectrometry. Since the compounds involved are very volatile and are capable of being admitted to a reservoir, a gas chromatographic or g.c.m.s. method may be more appropriate. Even though some of the components may not be resolved on the column, in a g.c.m.s. system they can be resolved mass spectrometrically so that the analysis can still proceed, as shown by examples in Chapter 6.

SAMPLE ADMISSION WITH FRACTIONATION

Reservoir systems

Where there is a substantial difference in the molecular weights of the components in the mixture, significant fractionation occurs during passage of the mixture through the leak between the reservoir and the source. In fact, the system can be envisaged as being a miniature version of the fritted glass type of molecular separator used in g.c.m.s. The flow rate of each component through the leak is inversely proportional to its molecular weight and directly proportional to the difference in its partial pressure across the leak. Thus, the spectra obtained immediately the sample is introduced into the reservoir reflect the presence of a greater proportion of the lighter components than the true proportion originally admitted to the reservoir. A straightforward calculation of the results as outlined earlier in this chapter (p. 94) can lead to a gross error in the calculated composition of the mixture. There are ways of reducing this error, however.

(a) The calculation can be carried out along the lines suggested previously (p. 94). Once the composition of the mixture entering the source has been calculated, a mixture can be made up of this composition and the spectrum determined. The spectrum of this new mixture will naturally differ from that of the original mixture because of fractionation, but will give an idea of the adjustment to be made in order to approximate more closely to the unknown mixture. The process can be repeated several times until a mixture is obtained which gives a mass spectrum acceptably close to that of the unknown.

This is undoubtedly a tedious process, and depends also upon obtaining spectra at a fixed time after admission to the reservoir, so that changes in composition of the various mixtures have occurred to the same extent in each case by the time the spectra are scanned. It is extremely difficult to achieve this because the time taken over the actual sample introduction inevitably differs from one run to the next. A way to surmount this problem is to have an on–off valve placed between the reservoir and the leak to the source. After sample introduction into the reservoir, the valve can be opened and the spectra run immediately.

(b) Another method of overcoming the problems of fractionation is to scan the spectra at certain fixed intervals of time such that there has been a significant change in the spectra during the interval. It is possible to extrapolate back to a spectrum at time zero. However, the usual reservoir size and porosity of the leak is such that a sample will last for several hours. In order to speed up the process smaller reservoirs, more porous leaks and a smaller sample level can be used. This will enable the change in composition to be followed more quickly to a point where the reservoir is virtually exhausted.

Direct inlet systems

The great advantage of a heated reservoir system over a direct inlet is in allowing an absolutely steady sample level to be maintained over a long period of time. On the other hand, direct inlet systems have the advantage that less volatile compounds can be determined. This is because the presence of the leak between

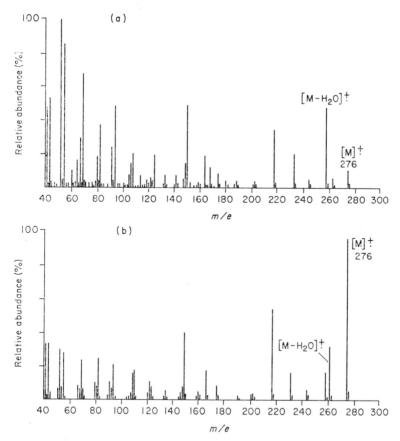

Fig. 5.2. The spectrum of 17β-hydroxyandrostane obtained on an MS-30 mass spectrometer: (a) via a heated reservoir; (b) via the direct inlet probe.

the reservoir and the source means that the pressure in the reservoir is higher than that in the source, whereas a direct inlet system operates at the source pressure, say 10^{-7} Torr. A compound will have to be heated to a higher temperature to enter a reservoir compared with introduction via the direct inlet, and therefore if thermally sensitive will almost certainly suffer degradation. The spectrum so obtained will not be the true spectrum of the compound. This fact is obvious if spectra recorded in the literature in the early 1960s prior to the widespread use of direct inlet systems are compared with those for the same compound recorded on a more modern mass spectrometer. To illustrate this, the spectrum of 17β-hydroxyandrostane obtained on an A.E.I. MS-30 both through the heated reservoir and the direct inlet are shown in Fig. 5.2.

The most obvious feature is that the molecular ion is greatly reduced in the case of the heated inlet spectrum, with the ion formed by loss of water considerably enhanced. This may be due to thermal loss of water in the reservoir prior to

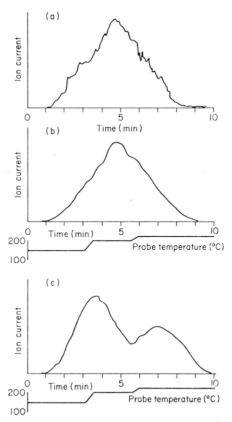

Fig. 5.3. (a) Evaporation profile of a compound evaporated from a probe heated only by the source. (b) Evaporation profile when an independently heated probe is temperature programmed. (c) Double peak occasionally obtained during temperature programmed evaporation. (Reproduced with permission from Ref. 1.)

ionization, but is more probably due to the extra internal energy of the molecules prior to ionization resulting in the molecular ion having a greater internal energy, so causing it to suffer more fragmentation.

The basic difficulty when using a direct inlet for quantitative work is that uneven evaporation of sample is obtained. If a pure sample is evaporated from a probe which is not heated independently, but depends for its heating upon the distance to which it is inserted into the heated source, a plot of total ion current would appear as in Fig. 5.3(a). An independently heated probe can provide a much smoother evaporation if time is taken to determine the optimum temperature programming. The total ion current profile recorded for the same compound evaporated from this type of probe would have the improved appearance shown in Fig. 5.3(b). However, there is sometimes a disadvantage in using temperature programming in that occasionally a double peak can be observed for the evaporation of a pure compound, giving the profile shown in Fig. 5.3(c).[1] This can be overcome by using the probe isothermally.

For several reasons it is desirable to achieve as smooth a profile for the total ion current as is possible when the sample is being evaporated from the probe. In the case of temperature programmed probes, therefore, some effort should be put into the determination of the best conditions necessary for a smooth output. One reason for doing this is that a jagged profile for the total ion current means that the sample pressure will also have varied fairly similarly during the evaporation. The work function of the filament will change slightly with the varying pressure, thus causing slight differences in the energy transferred to the molecules during ionization. This will then have some effect on the intensity of those fragment ions which are particularly sensitive to small differences in the internal energy of the molecular ions. The intensity of these fragment ions will not follow faithfully the total ion current profile in the case of a pure component, and for mixtures will not follow faithfully the changing sample level for that component. This effect is illustrated in Fig. 5.4, where a pure sample of N-methyltolbutamide

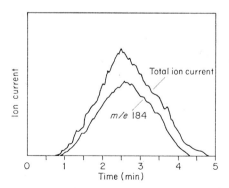

Fig. 5.4.   Evaporation profiles showing slight differences between total ion current and the intensity of the fragment ion m/e 184 for N-methyltolbutamide.

(**6**) has been rapidly evaporated from a direct inlet probe. It can be seen that the profile for the fragment ion m/e 184 is not quite the same as the total ion current profile.

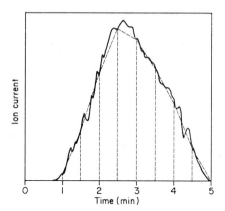

6

Another, more important, reason for requiring a smooth evaporation profile concerns the frequency of sampling of the mass spectrum. Unless selected ion monitoring is employed, which will provide the intensity of an ion over a period of time directly, it is usual to rely on retrieving the intensity of an ion or ions of interest from mass spectral scans, often over a restricted mass range. In many cases, the analysis depends upon reconstructing an area for the ion intensity versus time profile. The less smooth the actual evaporation, the greater is the error in the reconstructed evaporation profile as determined by the peak intensities during the scans. This is shown in exaggerated form in Fig. 5.5,

Fig. 5.5.   Comparison of real time evaporation profile for a fragment ion (solid line) and the profile reconstructed from repetitive scanning (broken line).

which depicts the actual profile for the evaporation of a pure sample and the reconstructed profile determined from the intensity of a selected ion in successive scans.

The situation can be improved by scanning at higher speeds and shorter intervals, so that the reconstructed profile approaches the real evaporation profile for that particular ion more closely. Where the sample level is low, this will cause an increasing error due to ion statistics (c.v. $= 100/\sqrt{N}$), since the faster the scan, the less the number of ions N collected at that particular m/e

value. For low sample levels therefore, it is particularly important to achieve as smooth an evaporation profile as is possible. Since this is not always practicable, a more feasible solution is to change the technique from scanning mass spectrometry to selected ion monitoring, thus increasing the dwell time for each m/e value to the maximum possible. As shown in Fig. 5.5, this has the result of producing a profile for the evaporation which is a replica of the real one, as well as increasing the sensitivity of the technique substantially.

Once the optimum conditions for the evaporation have been established, it is frequently possible to detect fine structure in the evaporation profile, due to the different evaporation behaviour of isomers.[2]

Possibly the least resolved cases are those concerning positional isomers of substituted aromatic compounds. Thus, in order to determine whether or not the o-, m- and p-tyramines, as their bisdansyl derivatives, underwent fractional

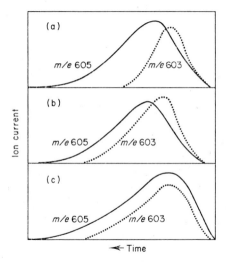

Fig. 5.6. Evaporation profiles for mixtures of the dansyl derivatives of $[^2H_2]p$-tyramine and (a) o-tyramine; (b) m-tyramine; (c) p-tyramine. (Reproduced with permission from Ref. 3.)

evaporation, it was necessary to label the *para* compound with deuterium to shift its molecular ion from m/e 603 to 605. Mixtures of $[^2H_2]p$-tyramine with o-, m- and p-tyramine gave the profiles shown in Fig. 5.6.[3] While the mixture of labelled and unlabelled p-tyramine, as expected, reached maxima at the same time, there was a small separation in the case of the other two isomers.

For fairly simple mixtures, the procedure to be followed for the analysis is analogous to that employed for the analysis of samples introduced into the mass spectrometer from a reservoir. It is necessary to choose ions in the spectra of the pure components such that the amount of cross-contribution is reduced to a minimum. One point to be borne in mind is that the repetition interval for the scan should be as short as possible, as discussed above (p. 102). This can be achieved by choosing ions at the high mass end of the spectrum in the case of downward scanning instruments, and ions at the low mass end of the spectrum for upward scanning instruments, the scan being terminated once all the ions of interest have been collected. Since background ions which cause interference decrease at higher m/e values, it is an advantage to have a downward scanning mass spectrometer so that ions of higher mass can be used for the analysis while still using a short scan.

As long as the interval between scans is kept constant, the area of the evaporation profile for each m/e value required for the analysis is given by adding together the intensities of response at that m/e value throughout the total number of scans. The mathematical treatment of these integrated responses is then the same as that for an n-component mixture.

The integrated ion current technique

Up to the present, quantitative mass spectrometry using a direct inlet probe is applied almost exclusively to the determination of compounds in complex biological mixtures. Many compounds of interest cannot be determined by g.c.m.s. since they are either thermally unstable and suffer thermal degradation at the g.c. injection port or on the column, or else they are too involatile to pass through the system. The direct inlet probe offers a means of quantifying such compounds down to a very low level without these problems of decomposition. Obviously, the simple treatment for an n-component mixture cannot be applied since n may not be known. More importantly, at low mass spectrometric resolving power, it may not be possible to choose an m/e value to which only the compound of interest contributes, since so many other components will be present, each having a large number of fragment ions in its spectrum. In general, the background from biological extracts is very high below about m/e 100 and decreases slowly with increasing m/e. Above about m/e 400 the background peaks become negligible in intensity, so that a derivative which shifts the molecular weight to something in excess of 400 is extremely useful. The probe technique has been applied particularly extensively to the determination of amines of biological interest, and 5-dimethylaminonaphthalene-1-sulphonyl chloride (dansyl chloride) has proved to be a useful derivatizing agent since it shifts the molecular weight considerably. Additionally, the derivatives are highly fluorescent, so that thin-layer chromatography (t.l.c.) can be utilized as a step to decrease the complexity of the mixture.

Another approach is to carry out the determination at high resolving power, so that the response at a given m/e value from the component of interest is separated from contributions due to interfering substances. The success of this method depends upon the component of interest having a molecular or fragment ion of rather unusual composition compared with the interfering substances, so that its mass is substantially different from the background. Drugs and metabolites tend to be amenable to the technique since they frequently contain elements not usually encountered in body fluids, such as fluorine, chlorine and sulphur. The dansyl or trifluoroacetyl, pentafluoropropionyl and heptafluoro-butyryl derivatives are also of use from this point of view.

The integrated ion current technique was first reported by Boulton and Majer[4-6] in 1970 as a method of determining *p*-tyramine (7) present in rat brain. This compound yields an ion at m/e 108 as follows:

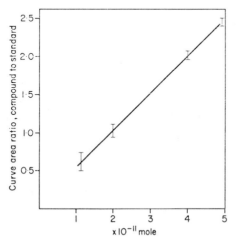

m/e 108

7

The spectrum of the extract at high resolving power showed that m/e 108 was a doublet, consisting of an ion $[C_8H_{10}]^+$ (108.0939) and the *p*-tyramine fragment ion $[C_7H_8O]^+$ (108.0575). The technique used was to tune in an ion from the reference compound perfluorotri-*n*-butylamine and set the ratio on the peak matching accessory to record m/e 108.0575. The switching arrangement in the peak matcher then swept through this ion every second or so, the output being recorded on the u.v. recorder. A calibration curve was constructed by evaporating known amounts from the probe and plotting the amounts versus the area under the curve of response versus time. Determinations in the range of 10^{-9} to 10^{-5} g were carried out with an error of $\pm 5\%$ by this method.

In this early work no internal standards were used, so the error involved in transferring known amounts of calibrating sample and extract would account for a considerable proportion of the observed error.

Fig. 5.7.   Calibration lines for the bisdansyl derivative of putrescine with the addition of 2.5×10^{-11} mole of the bisdansyl derivative of hexamethylene diamine as internal standard. (Reproduced with permission from Ref. 7.)

Seiler and Knödgen[7] investigated the use of the dansyl derivatives of the amine of interest and a chemically similar amine as an internal standard using a low resolving power of 1000. Thus, to measure putrescine as the bisdansyl derivative, a standard amount of 2.5×10^{-11} mole hexamethylene diamine bisdansyl derivative was added as internal standard. A linear calibration curve (Fig. 5.7) was obtained even in the presence of additional compounds in the mixture. Amounts down to 10^{-10} to 10^{-11} mole of these substances could be quantified with an error of $\pm 10\%$.

The presence of additional compounds other than the one being studied and its internal standard can cause a change in the evaporation profile of the two substances. When the compound and internal standard have very similar vapour pressures they evaporate from the probe at the same rate. A typical trace for a simple mixture is shown in Fig. 5.8(a), where m/e 318 from dansyl piperidine

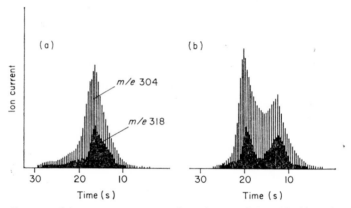

Fig. 5.8.   Integrated ion current curves of a mixture of 1×10^{-11} mole of dansyl pyrrolidine (m/e 318) and 2×10^{-11} mole of dansyl piperidine: (a) curve obtained without additional substances on the probe; (b) curves obtained with 10^{-9} mole of dansyl ethylamine on the probe. (Reproduced with permission from Ref. 7.)

and m/e 304 from dansyl pyrolidine are measured in a mixture of these compounds. The addition of a larger amount (about 100 times) of dansyl ethylamine has the effect of smearing out the evaporation profile and giving two maxima as shown in Fig. 5.8(b).

When applied to extracts, it was necessary to carry out some clean-up of the extract by a t.l.c. step, utilizing the fluorescent property of the dansyl derivative. The 5-di-n-butylaminonaphthalene-1-sulphonyl derivatives were said to be an improvement over the dansyl derivatives, leading to a threefold increase in sensitivity for the determination of piperidine in tissues.[8] The standard deviation for the determination of piperidine in the range 2–16 pmol when added to brain tissue was found to be $\pm 8\%$.

Problems can arise in the t.l.c. step when using chemically similar internal standards, since the standard will probably have a different R_f in the solvent system being used. Therefore, it is necessary to elute a rather wide band from the plate in order to include both of the compounds. This means that a significant amount of interfering substances will still be retained in the final extract. For this reason Boulton et al.[3] have used deuteriated analogues of the amines as internal standards on the grounds that they will have the same R_f values, thus reducing the amount of interfering substances eluted from the plate in the much narrower band. They favour the dansyl derivatives for amines, but unlike Seiler have used high resolving powers, of about 10 000, in order to reduce the interference which arises mainly from the solvent impurities. Besides its function in the t.l.c. step and for reducing errors due to losses in the various manipulations, the use of a deuteriated internal standard was found to be essential to provide a good calibration line. In the case of bisdansyl derivatives evaporated on their own from the probe tip, a plot of the logarithm of the area of the integrated ion

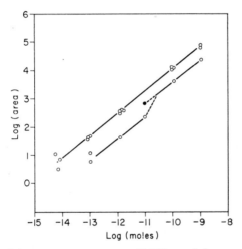

Fig. 5.9. Integrated ion current curves at 10 000 resolving power: upper: dansyl ethylamine, m/e 354.1402; lower: bisdansyl p-tyramine, m/e 603.1861. The solid circle indicates the signal produced by 10^{-11} mole of bisdansyl p-tyramine in the presence of 10^{-11} mole of the [$^2H_2$] analogue. (Reproduced with permission from Ref. 3.)

current profile versus the logarithm of the amount of compound gave the interesting calibration line shown in Fig. 5.9. Although a monodansyl derivative, that of phenylethylamine, gave a straight line, the bisdansyl derivative of p-tyramine did not give a linear response for amounts between 10^{-10} and 10^{-11} mole on the probe. This was true also for other bisdansyl derivatives, and is believed to be due to interaction between the dansyl amine molecules and the probe tip, causing decomposition with consequent lowering of the molecular ion intensity. A calculation of the amount of compound necessary to form a monolayer on the probe tip gave an approximate value of 2×10^{-11} mole. Thus, only this proportion of molecules on the probe tip could interact with the glass, allowing some of them to suffer decomposition. Different probe tips gave rise to different effects. Soft glass tips resulted in loss of response below about 10^{-10} mole, while silica tips behaved similarly to the borosilicate glass used originally. It is said to be particularly important to use a fresh tip for each determination, since silica tips, which are commonly reused after heating to redness in a flame, gradually acquired a salt coating which caused decomposition of small amounts of compound.

The use of a deuteriated internal standard overcame the problem of non-linearity of the calibration curve. Figure 5.9 shows that the response obtained for 10^{-11} mole of bisdansyl tyramine in the presence of an equal amount of bisdansyl [^{2}H$_2$]tyramine was sufficiently higher than the value for bisdansyl tyramine on its own to lie on the extrapolation of the linear portion of the calibration curve for the latter.

By this integrated ion current technique, the minimum detectable amount in the most favourable case (dansyl ethylamine) was found to be 2×10^{-15} mole, i.e. 0.7 pg at a resolving power of 7000.

The integrated ion current technique has been used recently for the determination of metalloporphyrins[9] present as contaminants in extracts containing small amounts of porphyrins. Porphyrins form complexes with high stability with traces of metal ions present in solvents, and this property has been used to analyse trace metals as tetraphenylporphyrin chelates by the integrated ion current technique.[10] In the work on contaminants, a gold crucible was used rather than a glass or silica probe tip, and about 25 to 30 scans were obtained during the evaporation of the sample, which took some 15 min. One difficulty in this type of work is the isotopic distribution in the molecular ion of metal containing compounds. This is illustrated in Fig. 5.10 for a scan through the molecular ion region of *copro*-porphyrin permethyl ester and its metal complexes.

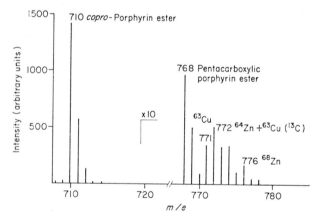

Fig. 5.10.   Scan through the molecular ion region of *copro*-porphyrin permethyl ester. (Reproduced with permission from Ref. 9.)

Before quantitative calibrations can be carried out for a case of this type, the corrected abundances of the molecular ions at m/e 710, 771 (^{63}Cu) and 776 (^{68}Zn) must be calculated. In a 500-ng sample, applied to the gold crucible in 10 μl of chloroform, it was possible to determine 0.5 ng of metalloporphyrin with an accuracy of 10%. These authors determined the response of various metallo*copro*-porphyrin permethyl esters relative to *copro*-porphyrin permethyl ester itself, so that all of these compounds could be determined from the same sample insertion.

It is naturally a great advantage to reduce the amount of manipulation required before a quantitative determination can be carried out. Taking this to the extreme, Snedden *et al.*[11-13] have monitored a number of compounds present in human tissue by simply placing 5-mg quantities of desiccated tissue on the probe and integrating the ion current from a series of diagnostic ions at high resolving power. In these early papers, compounds such as hypoxanthine,

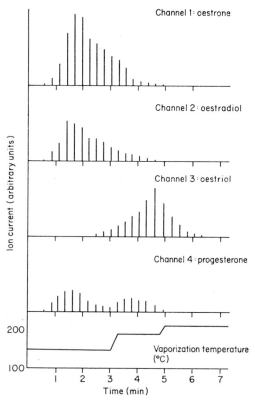

Fig. 5.11.   The evaporation profiles of a mixture of oestrone, oestradiol, oestriol and progesterone at 10 000 resolving power with a temperature programmed probe. Channel 1: m/e 270.162; Channel 2: m/e 272.178; Channel 3: m/e 288.173; Channel 4: m/e 314.225. (Reproduced with permission from Ref. 1.)

xanthine, uric acid and allopurinol were determined simultaneously with errors of the order of $\pm 20\%$ at the 10–100 ppm level.

By using high resolution selected ion monitoring at 10 000 resolving power, Snedden and Parker[1] were able to determine oestrone (m/e 270.162), oestradiol (272.178), oestriol (288.173) and progesterone (314.225) directly from dried tissue using a temperature programmed probe. The evaporation profiles for these compounds are shown in Fig. 5.11. The plots of profile area versus weight of steroid were linear down to 1 ng, the standard deviation for five replicates of 1 ng being $\pm 5\%$.

A simple approach such as this was also successful for determining the levels of the barbiturate heptabarbitone and its metabolites in urine without the need for extraction.[14] The molecular ions of these compounds are of low intensity, but the $[M - C_2H_5]^+$ ions are quite intense. These had accurate masses of 221.0926 (heptabarbitone), 235.0719 (3-ketoheptabarbitone) and 237.0875 (3- and 7-hydroxyheptabarbitone). The general procedure was to add 50 mg of an internal standard such as 2-chlorophenothiazine, which has a molecular ion at m/e 233.0666, to the total urine sample and then place 1-μl samples on the probe tip. A short scan to cover these ions was carried out repetitively at 10 000 resolving power, the intensities of the appropriate ions being summed from those scans that covered the period of evaporation of the compounds of interest. The ratio of the summed responses of each ion of interest relative to that for the internal standard were then interpolated on the calibration line. The value of high resolving power for separating ions due to the drugs, metabolites and internal standard from background contamination is shown in Fig. 5.12, where the m/e 221 region is shown from a urine sample containing heptabarbitone. Of the seven isobaric ions at m/e 221, the ion at m/e 221.0926, due to

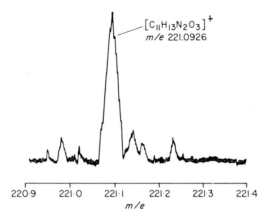

Fig. 5.12.   The m/e 221 region at 10 000 resolving power from a crude urine sample of a patient taking heptabarbitone. The sample was evaporated from a direct inlet probe. The ion at m/e 221.0926 is absent from a control urine sample, and therefore can be attributed to the $[M - C_2H_5]^+$ ion from heptabarbitone.

$[C_{11}H_{13}N_2O_3]^+$, is absent in a control urine sample, and therefore can be attributed entirely to $[M - C_2H_5]^+$ from heptabarbitone. In this work, it was possible to quantify 2 ng of these compounds on the probe with a coefficient of variation of 5%.

This wide range of applications illustrates the potential of the integrated ion current technique. It is slightly less convenient to perform than g.c.m.s. with selected ion monitoring for those compounds which would be amenable to both techniques, since care must be taken with the evaporation step. The measurement of the data is more time-consuming, since the response curves are not sym-metrical as in the case of g.c.m.s. where a peak height can suffice. In spite of this, it can be a more sensitive technique for high molecular weight polar compounds which suffer high losses both on the g.c. column and in the molecular separator, and may be the only sensitive method for those compounds not amenable to gas chromatography.

Chemical ionization

An alternative to using high resolving power as a method of reducing the contribution due to background contamination in the spectrum is to use chemical ionization instead of electron impact. This has the effect of increasing the proportion of the total ion current carried by the molecular or quasimolecular ions, thus improving the sensitivity of the analysis for those compounds (the majority) where no ion carries more than about 10% of the total ion current under e.i. conditions. This will also be true for the compounds causing the interference, but since the number of fragment ions from these compounds is also greatly reduced, the chance of a particular ion being formed with the same nominal mass as that being monitored will be much less. With the increased

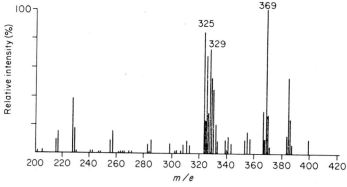

Fig. 5.13. Isobutane c.i. spectrum of a benzene extract from a plasma sample of a patient after administration of quinidine gluconate. The ion at m/e 325 is due to quinidine, that at m/e 329 from the $[^2H_2]$ internal standard and m/e 327 from dihydro-quinidine present in both the administered quinidine and the internal standard. The m/e 369 ion is probably derived from cholesterol by loss of H_2O from $[MH]^+$. (Repro-duced with permission from Ref. 15.)

availability of c.i. sources, several laboratories have now employed direct insertion analyses under c.i. conditions.

The fairly clean background in an isobutane c.i. direct inlet spectrum of a benzene extract from a plasma sample is shown in Fig. 5.13.[15]

To the plasma had been added 1.17 μg of [^{2}H$_2$]dihydroquinidine as internal standard for the dihydroquinidine present in the quinidine administered to the patient. The base peak in the spectrum at m/e 369 is derived from the quasi-molecular ion of cholesterol by the loss of water. In this work, the object was to investigate the feasibility of isobutane c.i. mass spectrometry for the quantification of quinidine, lidocaine and a metabolite of lidocaine, monoethylglycine-xylidide in plasma by using deuteriated analogues of these compounds as internal standards. The assumption was made that the deuteriated compounds had evaporation characteristics identical to those of the unlabelled compounds, so that it was only necessary to carry out one scan of the spectrum. The range of detection was from 5 ng to 4 μg per ml of plasma for lidocaine, 0.1 μg to 1 μg per ml for its metabolite and 1 μg per ml for quinidine.

Isobutane c.i. mass spectrometry has also been used for a quantitative assay of amino acids in dried blood spots.[16] Deuterium or ^{15}N labelled amino acids were used as internal standards. A single step procedure converted all the amino acids into more volatile N-acetyl methyl esters, and the isobutane spectra of these yielded [M + 1]$^+$ ions. Except for leucine and isoleucine, each amino acid has a unique [M + 1]$^+$ ion. A simple temperature programme was used in

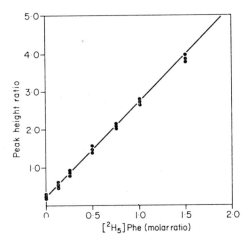

Fig. 5.14. A calibration line for phenylalanine. Whole blood was supplemented with phenylalanine to produce concentrations of 5, 10 and 15 nmole per 5 μl of dried blood. A fixed amount of 10 nmole of [^{2}H$_5$]phenylalanine was used for each analysis. The peak height ratios of m/e 222 to m/e 227 from unlabelled and labelled phenylalanine were obtained from the scanned spectra. The labelled phenylalanine contained species from [^{2}H$_0$] to [^{2}H$_7$], the amounts of [^{2}H$_0$] and [^{2}H$_5$] being 4 and 37% respectively. (Reproduced with permission from Ref. 16.)

which spectra were obtained at 50 °C intervals from 50 up to 300 °C, thus yielding five spectra. Calibration curves were prepared by averaging the ratio of the responses for labelled and unlabelled amino acid derivative from these five spectra. Figure 5.14 shows a typical calibration curve for phenylalanine. This was obtained by adding 10 nmole of labelled phenylalanine to blood spiked with 5, 10 and 15 nmol of phenylalanine per 5 μl of dried blood spot.

By this technique it was possible to determine all the common amino acids down to 25 ng with a signal-to-noise ratio of better than 10:1. It is interesting that mass spectrometry gave results very close to those produced by ion exchange chromatography or an amino acid analyser for the determination of phenyl-alanine in blood and plasma from patients suffering from phenylketonuria, as shown in Table 5.2.

TABLE 5.2
Comparison of mass spectrometry and ion exchange chromatography for the determination of phenylalanine in phenyl-ketonuric blood and plasma

Patient	Mass spectrometry, blood spot concentration (mg %)[a]	Ion exchange, plasma[b]
1	18.48 ± 0.30	18.81
2	20.60 ± 0.25	20.30
3	14.98 ± 0.29	15.80
4	13.09 ± 0.24	13.04

[a] Mean of five determinations ± standard error of the mean.
[b] Mean of two determinations.

An interesting study compared the results obtained using a direct inlet probe and g.c.m.s., both with methane c.i., for the quantification of tolbutamide and its metabolites.[17] Unfortunately, the N-1 methyl derivatives of tolbutamide (8) and its metabolites (9 and 10) undergo partial decomposition under g.c. conditions to yield the N-1 methylated sulphonamides, so that it was felt that g.c.m.s. might not be the ideal technique for the analysis.

CH_3—⬡—$SO_2\overset{\underset{\displaystyle CH_3}{|}}{N}$—CO—NH C_4H_9

8

$HOCH_2$—⬡—$SO_2\overset{\underset{\displaystyle CH_3}{|}}{N}$—CO—NH$C_4H_9$

9

CH_3O—$\overset{\overset{\displaystyle O}{||}}{C}$—⬡—$SO_2\overset{\underset{\displaystyle CH_3}{|}}{N}$—CO—NH C_4H_9

10

CH_3—⬡—$SO_2\overset{\underset{\displaystyle CH_3}{|}}{N}$—CO—NH$CD_2CH_2CH_2CH_3$

11

In both sets of experiments deuteriated internal standards were used, with a computer system acquiring data from selected ion recording during evaporation from the probe or elution from the gas chromatograph. The probe was heated from 50 to 150 °C, and extremely good evaporation profiles were obtained. The computer reconstructed profile for the $[M + 1]^+$ ions from N-1 methyl-tolbutamide at m/e 285 and its dideuteriated analogue (11) at m/e 287 is shown in Fig. 5.15.

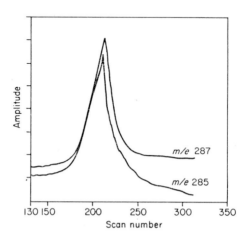

Fig. 5.15.   The computer reconstructed evaporation profile of N-1-methyltolbutamide (m/e 285) and N-1-methyl-[^{2}H$_2$]tolbutamide (m/e 287). (Reproduced with permission from Ref. 17.)

The computer was programmed to calculate the areas of the profiles, and the levels of these compounds in plasma obtained by interpolation on the calibration curve. At a level of 25 μg per ml of tolbutamide, the c.v. was 3.6% for the probe determination and 3.8% for the g.c. method. The sensitivity of the probe method is illustrated by the fact that levels of 200 ng per ml in plasma were determined without difficulty.[18]

In the tolbutamide work, the background contribution from blank plasma extracts was small and could be subtracted from the response obtained from drug containing plasma.

Although at present the majority of instruments operating under c.i. conditions are of low resolving power, these sources can equally well be used on high resolution instruments. This offers the possibility of carrying out the c.i. direct inlet determinations at high resolving power, giving the advantages of both methods discussed above for overcoming the problem of interference from endogenous substances in biological extracts.

REFERENCES

1. W. Snedden and R. B. Parker, *Biomed. Mass Spectrom.* **3**, 295 (1976).
2. J. R. Majer and R. Perry, *J. Chem. Soc. A* 822 (1970).
3. D. A. Durden, B. A. Davis and A. A. Boulton, *Biomed. Mass Spectrom.* **1**, 83 (1974).
4. A. A. Boulton and J. R. Majer, *J. Chromatogr.* **48**, 322 (1970).
5. A. A. Boulton and J. R. Majer, *Nature (London)* **225**, 658 (1970).
6. A. A. Boulton and J. R. Majer, *Can. J. Biochem.* **49**, 993 (1971).
7. N. Seiler and B. Knödgen, *Org. Mass Spectrom.* **7**, 97 (1973).
8. N. Seiler and H. H. Schneider, *Biomed. Mass Spectrom.* **1**, 381 (1974).
9. E. Larsen, H. Egsgaard, J. Moller and T. K. With, *Biomed. Mass Spectrom.* **4**, 113 (1977).
10. B. A. Davis, K. S. Hui, D. A. Durden and A. A. Boulton, *Biomed. Mass Spectrom.* **3**, 72 (1976).
11. R. B. Parker, W. Snedden and R. W. E. Watts, *Biochem. J.* **115**, 103 (1969).
12. R. B. Parker, W. Snedden and R. W. E. Watts, *Biochem. J.* **116**, 317 (1970).
13. W. Snedden and R. B. Parker, *Anal. Chem.* **43**, 1651 (1971).
14. B. J. Millard, M. G. Lee and N. J. Haskins, *Advances in Mass Spectrometry*, Vol. 6 (A. R. West, editor), 1974, p. 181.
15. W. A. Garland, W. F. Trager and S. D. Nelson, *Biomed. Mass Spectrom.* **1**, 124 (1974).
16. J. M. L. Mee, J. Korth, B. Halpern and L. B. James, *Biomed. Mass Spectrom.* **4**, 178 (1977).
17. S. B. Matin and J. B. Knight, *Biomed. Mass Spectrom.* **1**, 322 (1974).
18. J. B. Knight and S. B. Matin, *Anal. Lett.* **7**, 529 (1974).

CHAPTER 6

QUANTIFICATION IN GAS CHROMATOGRAPHY MASS SPECTROMETRY SYSTEMS

Two methods are at present used to quantify compounds introduced into the mass spectrometer via a gas chromatograph. The first, and most popular one is to monitor the intensity of a limited number of ions which are characteristic of the compounds being determined, by switching the mass spectrometer analyser rapidly between these ions. The technique of focusing the mass spectrometer on just one ion generated from the effluent of a gas chromatograph was first used by Henneberg[1,2] and Henneberg and Schomburg[3] in a study of hydrocarbons, alcohols and esters, while Sweeley et al.[4] introduced the concept of focusing on several ions simultaneously. The second method is to scan the spectrum repetitively, with the data from these scans being acquired by a computer. As shown by Hites and Biemann,[5] the varying intensity of selected ions can be retrieved from the scans, and the output presented in a similar fashion to the real time output of the first method.

There are several important differences between these two techniques which warrant two distinct terms to describe them. The first method is much more sensitive than the second, by up to a factor of several thousand, depending upon the sampling time, but requires a foreknowledge of which ions are needed for the analysis. The second, while less sensitive, has the great advantage that any ion or ions of choice can be retrieved from the scan data at the termination of the experiment, so that the most useful ions for the analysis can be determined without the need for further injections of sample.

The mention of the confused state of the literature refers to the fact that at least 13 terms have been used for the method of real time monitoring, with some of these terms also being applied to repetitive scanning.[6] The original term 'mass fragmentography' coined by some early workers in the field obviously causes difficulty for those who use c.i. mass spectrometry, since such workers may never monitor a fragment ion. Mass spectrometry journals are becoming more or less standardized on 'selected ion monitoring' as a phrase which conveys the basic concept of the technique whereby the mass spectrometer is caused to switch rapidly between a limited number of ions, and this term will be used here. For the method by which ions are selected from scans that have been acquired previously, the term 'selected ion retrieval' will be used, since it is considered that the phrase is self-explanatory to those with the most elementary knowledge of mass spectrometry. It is not necessary for the scans to be retrieved by a data system; the measurements could also be made on chart spectra produced from a limited number of repetitive scans.

SCANNED SPECTRA

In Chapter 5, the scan rate of the mass spectrometer was not of prime importance, since sample levels introduced from non-chromatographic inlet systems remain reasonably constant over the course of a few minutes, although some compounds cause difficulty on the probe. In the case of peaks eluting from the gas chromatograph, the scan speed of the mass spectrometer is of crucial importance. For packed columns, an average time across a g.c. peak can be taken as being 20 s. The scan speeds of 10 or 30 s per decade in mass which are typical for probe spectra would lead to severe distortion of the intensities of the ions in the spectra, since the sample level could change by a factor of 10 during the mass spectral scan. In order to decrease this distortion of intensities, the scan speed should be such that about six spectra can be obtained during elution of the g.c. peak. This means that scan speeds of 1–3 s per decade are necessary.

For an exponential scan, the time spent on a mass spectral peak between the points at which the intensity is 5% of that at the maximum is given by

$$t = \frac{0.43t_{10}}{R}$$

where t_{10} = s per decade for the scan, and R = static resolving power. The time t is the same for all the peaks in the spectrum, this being a fundamental property of exponential scans.

At high scan speeds and high resolving power, the time spent on a peak becomes small, perhaps less than a millisecond, and this places considerable demands on the amplifying and recording system. For the MS-30 mass spectrometer the amplifier bandwidths can be set automatically to 100, 300, 1000, 3000, 10 000 or 30 000 Hz by the switch that selects the scan speed at particular resolving powers. Table 6.1 gives the time spent on a peak for different scan speeds and resolving powers, and the appropriate bandwidth settings. The minimum bandwidth necessary for particular scan speeds and resolving powers can be determined by the approximate equation

$$B \approx \frac{7.3R}{t_{10}}$$

The response of the amplifier is not fast enough to cope with a scan speed of 1 s per decade at 10 000 resolving power. An important limiting factor becomes the response of the galvanometers in the recorder. The galvanometers typically have a frequency response of 2000 or 5000 Hz. From the data in Table 6.1, such

TABLE 6.1
Time spent on a peak and the minimum amplifier and recorder frequency response necessary at different resolving powers and scan speeds for exponential scans

Scan speed (s per decade)	Resolving power	Time spent on peak (ms)	Minimum amplifier and recorder response necessary (Hz)
1	1000	0.43	10000
	1500	0.286	10000
	3000	0.143	30000
	10000	0.043	100000
3	1000	1.29	3000
	1500	0.858	3000
	3000	0.429	10000
	10000	0.129	30000
10	1000	4.3	1000
	1500	2.86	1000
	3000	1.43	3000
	10000	0.43	10000
30	1000	12.9	300
	1500	8.58	300
	3000	4.29	1000
	10000	1.29	3000

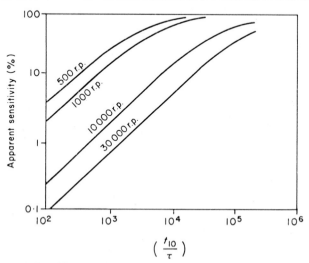

Fig. 6.1. The relationship between the sensitivity and t_{10}/τ during scanning of a spectrum at a number of resolving powers (r.p.): t_{10} = scan time in s per decade in mass, and τ = frequency response of amplifying and recording system. (Reproduced with permission from Refs. 7 and 8.)

a galvanometer recorder cannot be used for scan speeds of 3 s per decade, which is necessary for g.c. work, unless the resolving power is kept below 1500.

For an amplifying system which uses a galvanometer recorder, Banner has studied the relationship between sensitivity and scan rate to frequency response ratio for a number of resolving powers.[7,8] The results are presented in Fig. 6.1.

Table 6.1 showed that a bandwidth of 10 000 was necessary for a 1 s per decade scan at a resolving power of about 2000. Reference to Fig. 6.1 shows that if the bandwidth is 1000 Hz at 1 s per decade, $t_{10}/\tau = 10^3$ and the apparent sensitivity is only about 10 % of that for static operation, i.e. when the mass spectrometer is focused constantly on one m/e value. The visual effect of scanning at successively faster rates with insufficient bandwidth for three adjacent peaks in a spectrum is shown in Fig. 6.2.

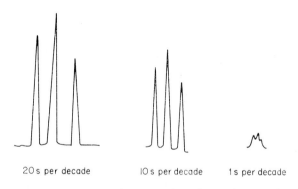

| 20 s per decade | 10 s per decade | 1 s per decade |

Fig. 6.2.   Effect of scanning three adjacent peaks at increasing speeds with insufficient bandwidth.

It will be noticed that not only are the intensities of the peaks lowered for faster scans, but that the valley between the peaks, i.e. the apparent resolution, is also degraded. Banner showed that the dynamic resolving power of the mass spectrometer depends upon the scan rate to time constant ratio according to the curves in Fig. 6.3. Taking the situation as before of a scan at 1 s per decade and a system bandwidth of 1000 Hz, where $t_{10}/\tau = 10^3$ the apparent resolving power is 100, irrespective of whether the static resolution is 500 or 10 000.

For faster scan rates at higher resolving powers where a galvanometer recorder cannot be used, a data system with a fast digitization rate of at least 50 000 Hz is necessary. At a resolution of 3000 and a scan speed of 1 s per decade, Table 6.1 showed that only 0.143 ms are spent in traversing a peak between the 5 % intensity points. Even with a 50 000 Hz digitization rate, only seven samples would be taken across the peak, which is inadequate to define its shape properly. As a general rule, scan speeds of 1 s per decade can be used only at resolving powers of 1000 or less.

For the linear scans produced by the quadrupole mass spectrometer, the time spent on a peak depends upon the starting point, and the resolution changes

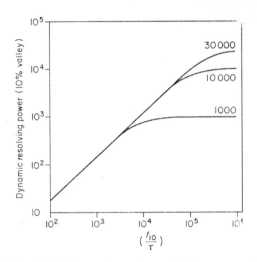

Fig. 6.3. The relationship between dynamic resolving power and t_{10}/τ for various static resolving powers of the mass spectrometer: t_{10} = scan time in s per decade in mass, and τ = frequency response of amplifying and recording system. (Reproduced with permission from Refs. 7 and 8.)

along the scan. If the decade from m/e 50 to 500 is scanned in 1 s, the time spent on a peak is $1/450 = 2.2$ ms. The above comments on loss of sensitivity in the case of magnetic instruments still apply for quadrupoles, so that the best approach is to calculate the time spent on a peak from a knowledge of the scan speed and the starting and finishing m/e values in order to determine the characteristics necessary for the recording and amplifying system.

SELECTED ION MONITORING

Relative sensitivity of selected ion monitoring and retrieval

Since selected ion retrieval is carried out from repetitively scanned spectra, the relative sensitivities of the two techniques of selected ion monitoring and selected ion retrieval are a reflection of the response obtained by constantly monitoring a peak, and the response obtained when the peak is scanned in a mass spectrum. These relative sensitivities depend upon the sample level, since at very low sample levels the number of ions collected in a mass spectral peak scanned at high speed or at high resolving power becomes small enough for ion statistics to be the limiting factor.

High sample levels

The previous section covered the effect of insufficient bandwidth upon the response obtained for a peak when sample levels were high enough for ion statistics to have virtually no influence. Provided that the bandwidth is sufficient for the scan conditions used, the loss of response is negligible for such situation. This is shown by Fig. 6.4, where the response for continuous monitoring of a mass spectral peak is compared with the responses during scans of 100 s per decade and 1 s per decade. A statement of sensitivity should include a mention of the signal-to-noise (S/N) ratio, otherwise it becomes meaningless. Inspection of Fig. 6.4 shows that the S/N ratio is much worse for the scan at 1 s per decade than the scan at 100 s per decade. This is because a lower bandwidth can be used for the slower scans. When an S/N ratio is greater than 1, it is improved by reducing the bandwidth, the improvement being inversely proportional to the square root of the bandwidth. With the bandwidths for the scans at 1 and 100 s per decade being 100 Hz and 10 000 Hz respectively, this means that there is a tenfold increase in the S/N ratio by scanning at the slower speed.

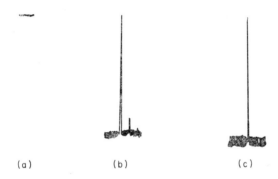

(a)   (b)   (c)

Fig. 6.4.   (a) Response level when the mass spectrometer is tuned to the top of a peak. (b) The same peak scanned at 100 s per decade. (c) The same peak scanned at 1 s per decade. The signal-to-noise ratio in (b) for the scanned peak is about 10 times that for (c) on the logarithmic output.

For continuous monitoring of a mass spectral peak in a compound which is eluting over a period of, say, 20 s, the bandwidth can be reduced to 1 or 2 Hz with an even greater improvement in the S/N ratio compared with the fast 1 s per decade scan.

Although sharp peaks emerging during the early portion of a gas chromatographic run may well require scan speeds of 1 s per decade, the broader peaks of, say, 20–30 s width which emerge later in the run on a packed column can be covered readily by a slower scan of 3 s per decade, while later in the run even 10 s per decade may suffice. This points to the advantage which could be obtained by gradually increasing the seconds per decade time throughout the period of the g.c. run, since much better quality spectra will be obtained on the later g.c. peaks. With single beam instruments which are set up initially by external calibration on PFK for the data acquisition system using the times of arrival of the peaks, a correction would be necessary for the computer to locate the correct m/e values. This may be easier to perform on a double beam instrument where the PFK is constantly available in the source, or on systems where mass assignment is based on a Hall probe.

A further improvement in scanned spectra may be obtained by limiting the range in mass over which the spectra are scanned. For, example, if only half a decade in mass is covered, say from m/e 500 down to m/e 100, the scan time can be doubled and the bandwidth halved. This will improve the S/N ratio by a factor of $\sqrt{2}$ compared with a full decade scan.

Under conditions of high sample level, the S/N ratios for all except the weakest peaks in the spectrum are so high that a 10- or even 100-fold improvement becomes meaningless. Because of the greater flexibility of selected ion retrieval, where components of interest can be injected onto the gas chromatograph at the tens of nanograms level, it should be preferred to the use of selected ion monitoring.

Low sample levels

The arrival of an ion at the first dynode of the electron multiplier is multiplied by secondary emission to form an avalanche of up to 10^6 electrons from the final dynode, depending upon the gain. These electrons flow through the load resistance of perhaps 10^7 Ω to give rise to a voltage impulse. If the response time of the circuit is, say, 10^{-4} s, the voltage amplitude V is given by

$$V = \frac{\text{electronic charge} \times \text{multiplier gain} \times \text{load resistance}}{\text{response time}}$$

$$= \frac{1.6 \times 10^{-19} \times 10^6 \times 10^7}{10^{-4}} = 16\,\text{mV}$$

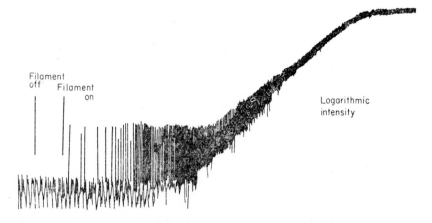

Fig. 6.5.   The change in the signal level and noise as an ion is tuned in by adjusting the magnet current. The display is on a logarithmic intensity scale. Before the filament is switched on, the noise is due to the multiplier and amplifying system. The large spikes obtained once the filament is switched on are due to the arrival of single ions.

These pulses are easily discernible on an oscilloscope and on a galvanometer recorder if the bandwidth of the amplifier is kept high.

The arrival of single ions at the multiplier at such a frequency that they do not overlap, so that single impulses are observed, can be considered to be noise, since no information can be obtained from such a signal. Once the rate of arrival of single ions increases so that the impulses overlap, a meaningful signal is obtained. The transformation between these two states can be seen in Fig. 6.5. The trace was obtained by tuning the mass spectrometer to a point some way from a peak and turning the multiplier gain to about 10^6. The bandwidth of the amplifier was kept at 10 kHz so that single ions could be seen readily at the fast chart speed of 500 mm per second. While the chart was running the filament was turned on and the peak was tuned in slowly to its maximum intensity. In order to improve the dynamic range of the recorder, a logarithmic intensity output is used. This means that a signal which appears twice as intense as another signal is actually 10 times as intense. Some interesting points emerge from a consideration of the resulting trace.

The noise level prior to turning on the filament is low and of much lower frequency than the noise due to single ions arriving at the collector once the filament is turned on. The S/N ratio for each of the single ion pulses is of the order of 50:1, based on the (logarithmic) ratio of the average height of the single ion pulse to the lower frequency noise. Once these single ions start to arrive in rapid succession, they constitute a new noise level which is 50 times worse than that before the filament was turned on. As the arrival of ions increases still further so that the impulses become superimposed, the gross fluctuations of the single ions is exchanged for the smaller fluctuations due to slight differences in the rate of arrival of ions, i.e. fluctuation due to ion statistics. Since the variance depends upon the reciprocal of the square root of the number of ions, this fluctuation decreases as the number of ions reaching the collector increases. Thus, the noise at the top of the peak is reduced to a very low level compared with the noise at the base of the peak where single ions arrive separately.

Under typical conditions on the MS-30 instrument in this laboratory, the height H of a peak obtained during a scan is given by

$$H \approx \frac{nh}{7}$$

where n = number of ions in the peak, and h = height of single ion pulse.

This means that the height of a mass spectral peak which contains less than about seven ions is no more than the height of a single ion pulse. This reflects the chances that such a small number of ions arrive separately across the profile of the peak, rather than that two or more arrive simultaneously. With present-day mass spectrometers the S/N ratio for a single ion may be perhaps 50:1. Thus, for a seven ion peak the S/N ratio is still 50:1. Although the peak is no more intense than a single ion peak, it will be wider, and hence more significantly a mass spectral peak.

If it is assumed that the peak is triangular, it will contain only half of the ions that would otherwise be collected during the same time under static conditions, so the height H of the peak is given by

$$H \approx \frac{\text{number of ions at that } m/e \text{ value entering the magnetic analyser}}{14} \times h$$

For a particular sample whose molecular weight and mass spectrum are known, the minimum amount of that sample entering the source which will yield 14 ions during the width of the mass spectral peak can be calculated.

The total number of ions produced by a compound at a particular m/e value is given by

$$\frac{6.023 \times 10^{23} \times 10^{-3} \times \%\Sigma \times w}{\text{molecular weight}}$$

where 10^{-3} represents the ionization efficiency, which is typically 0.1%, $\%\Sigma$ is the percentage of the total ion current at that particular m/e value, and $w =$ weight of the compound.

If the g.c. peak in which this number of ions elute has a width of T seconds, and the mass spectral peak at the particular scan speed and resolving power used is t seconds wide, the number of ions n produced during the time for which the mass spectral peak is scanned is given by

$$n = \frac{2t}{T} \times \text{total number of ions at that } m/e \text{ value}$$

This makes the assumption that the particular m/e value is scanned just as the g.c. peak is maximizing. Thus,

$$n = \frac{2t \times 6.023 \times 10^{23} \times 10^{-3} \times \%\Sigma \times w}{T \times \text{molecular weight}}$$

Some typical values can be put in the equation as follows: scan speed is 3 s per decade at 1000 resolution, so $t = 1.29 \times 10^{-3}$ s, g.c. peak width is 30 s, molecular weight is 500, and $\%\Sigma$ is 1%, i.e. 0.01. Then

$$n = \frac{2 \times 1.29 \times 10^{-3} \times 6.023 \times 10^{23} \times 10^{-3} \times 0.01 \times w}{30 \times 500} = 1.035 \times 10^{12}w$$

In order to give a peak height in the mass spectrum which is larger than the pulse for a single ion, as noted above, n must be at least 14:

$$w = \frac{14}{1.035 \times 10^{12}} = \text{approximately 13 pg}$$

This makes no allowance for the efficiency of the molecular separator, which is typically 50%.

Under these conditions, it requires, say, 25 pg injected on the column, and the good fortune for the scan to be passing through the m/e value of interest just as the g.c. peak maximizes, before a significant peak is observed at an m/e value which represents 1% of the total ion current.

For a peak which represents 10% of the total ion current, the corresponding amount would be about 2–3 pg. An idea of the vastly greater sensitivity of selected ion monitoring can be gained from the fact that a recent paper reports the measurement of prostaglandin $F_{2\alpha}$ at the 50 fg level by single ion monitoring, with an S/N ratio of about 50:1.[9]

These authors have concluded that the lower detection limit for prostaglandin $F_{2\alpha}$ as the methyl ester tri-TMS derivative is about 10 fg. Since the molecular weight is around 500, this represents approximately 10^7 molecules. If the ionization efficiency is taken to be 0.1%, and the $\%\Sigma$ for the m/e 423 ion is not more than 10%, the number of ions monitored during elution of the g.c. peak representing this substance is of the order of 1000. For the 50 fg level, the precision of the measurement was about 11%. If, for purposes of comparison, it is assumed that the variance was due totally to the mass spectrometric response rather than the g.c. injections, it is necessary to calculate how much compound would give the same precision when the spectrum is scanned at 3 s per decade at 1000 resolving power. Since the c.v. $= 100/\sqrt{n}$, where $n =$ number of ions in the peak, this number n needs to be about 82. In the previous calculation for scanned spectra it took 13 pg to yield 14 ions, so that about 80 pg are required to produce the same coefficient of variation in the scanned spectra as 50 fg produced in single ion monitoring. On this basis, the relative sensitivities of the two techniques are in the ratio of about 1600:1.

For the reasons discussed earlier (p. 122), when a high sample level is used, there is no significant difference between the sensitivity of single ion monitoring and selected ion monitoring in which a limited number, say four, channels are monitored. However, at the low levels discussed in this section, there will be a difference due mainly to ion statistics. The number of ions collected per channel in an N channel selected ion monitor is of course n/N, where n would be the number of ions collected by single ion monitoring. At low levels the sensitivity will be proportional to the number of ions collected, so that the N channel monitor will be only $1/N$ times as sensitive as single ion monitoring. As noted earlier, the c.v. due to ion statistics $= 100/\sqrt{n}$. Thus the c.v. for N channels is N times as high as the c.v. for single ion monitoring. For a four-channel monitor, the c.v. will be twice as high as for one channel which is not switched. This excludes any time spent not dwelling at the particular mass values being monitored; dead time of this nature would make the comparison even worse. This difference in the coefficients of variation is an important consideration for those cases towards the limits of detection where n may be less than, say, 1000 ions.

Eyem[10] has studied the relationship between dwell time spent on a peak, the ion flow during the elution of the g.c. peak and the c.v. obtained. The ion flow fluctuations superimposed on the concentration changes of the substance should be less than or equal to the desired signal height measurement precision at the apex of the peak. Table 6.2 shows the ion flow per second necessary for various dwell times in order to achieve coefficients of variation of 1, 5 and 25%.

TABLE 6.2
Relationship between ion flow per second, dwell time and coefficient of variation

Coefficient of variation (%)	Dwell time (ms)			
	50	250	1250	6250
1	200000	40000	8000	1600
5	8000	1600	320	64
25	320	64	12.8	2.56

By considering the Gaussian curve of a g.c. peak of, say, 20 s width, if the desired precision of the signal height is to be 2%, the ion flow to the electron multiplier of 1.4×10^4 ions per second should be essentially constant for a time of 180 ms either side of the g.c. peak maximum.

An application of computer control to mass spectrometers is now emerging in which the advantages of selected ion monitoring are applied to the whole spectrum. The major disadvantage of a scanned spectrum is that for the majority of the scan time the mass spectrometer is not passing through a peak position, but is somewhere between adjacent peaks, or is passing through a peak position where no peak is formed from the compound. Since the voltages on the rods of a quadrupole mass spectrometer can be switched rapidly and conveniently to the values necessary to focus ions as an alternative to changing in a smooth ramp in the usual scanning mode, these make ideal instruments for total spectrum monitoring. By this is meant a system in which the instrument is switched to one m/e value for a certain time, then jumps to the next m/e position for the same length of time, the process being repeated until the whole spectrum has been acquired. A partial system was reported by Green and Hertel,[11] who monitored some 15 ions sequentially. However, they were concerned with rapid compound identification rather than quantification. Fifteen m/e values were considered to give an unequivocal identification of drugs when a crude mixture was flash evaporated into the source via a membrane separator, with no chromatography being carried out, but some false positives have been obtained. The advantage of the method was the rapidity with which such a contracted spectrum could be obtained.

A total spectrum monitoring instrument has been described by Reynolds[12] in which, for example, 8 ms is spent acquiring ions at each m/e value, with a 2 ms switching time to the next m/e position. The signal at each mass position becomes effectively stationary, therefore, as opposed to the time dependent signal obtained during a normal scan. A full integrator can then be used on the signal to improve the S/N ratio. The system is substantially more sensitive than a conventional scan, as shown in Table 6.3. The time constant of the amplifiers

TABLE 6.3
Comparison of attributes affecting S/N efficiencies. Scanning instruments from m/e 50 to 500 in 4500 ms

Type of scan	Time on peak (ms)	Time constant of amplifiers (ms)	Ions detected (n ions per ms at peak top)	Effective noise bandwidth (Hz)	$1000 \times$ detected ions $n \times$ bandwidth
Control and integrate	8	N/A	$8\,n$	40	200
Linear $(m = m_0 + k_1 t)$	10	2	$5.1\,n$	80	64
Parabolic $(m = m_0 + k_2 t^2)$	6.6	1.3	$3.4\,n$	120	28
Exponential $(m = m_0 e^{-k_3 t})$	3.9	0.8	$2.0\,n$	200	10

is chosen to be 0.2 times the peak width. This degrades the resolution by some 25–35%, but allows a comparison of the systems. For normal instruments, the ions detected in a Gaussian shaped peak are given by 0.51 × time spent on peak × ion flow per millisecond. For the controlled quadrupole instrument, the ions detected will be equal to the time spent on the peak × ion flow per millisecond, since the instrument is always centred on the peak. The final column in the table gives a comparison between the four systems of computer controlled, linear scan, parabolic scan and exponential scan, showing that the computer controlled quadrupole is the best of the four systems.

Specificity of selected ion monitoring

The mass spectrometer can be considered to be the ultimate in specific detectors, since, with a few exceptions, a mass spectrum is completely unique. At low resolving power, where no distinction can be made between different elemental compositions at the same integral m/e value, the uniqueness of the spectrum can be preserved by surprisingly few peaks. This is because the intensities of the peaks can be taken into account. If even a very crude intensity scale is used where a peak is represented by the numbers from 1 to 10, i.e. 0–5% being 1, 6–10% being 2, etc., the number of possible combinations for a mass spectrum with n peak positions is 10^n. Thus, in theory, eight channels with intensity measurements could represent a hundred million compounds. In a typical problem, there would not be any need to cover such a large number of possibilities. In a complex biological mixture, for example, there are unlikely to be more than a few thousand compounds. Such a small number could be distinguished by four channels, since with intensity measurements on the scale 1 to 10 this would give 10^4 different combinations.

The situation is not quite so straightforward, however. A spectrum is not limited to only four channels. Four ions can be chosen from a mass scale extending to the limit of the instrument, say 750 mass units. Four mass positions can be chosen in $750 \times 749 \times 748 \times 747$ ways, which is approximately equal to 3.1×10^{11} ways. This leads to the conclusion that four ions chosen at random from a mass scale going up to m/e 750 can cover more compounds than are ever likely to be synthesized or encountered in extracts. Table 6.4 examines the

TABLE 6.4
Number of compounds that can be represented by a limited number of channels chosen from 750 m/e values. Intensities can be allocated to one of 10 values

Number of ions monitored	Number of ways of choosing from 750 m/e values	Number of different compounds that can be represented
1	750	750
2	5.6×10^5	5.6×10^7
3	4.2×10^8	4.2×10^{11}
4	3.1×10^{11}	3.1×10^{15}

different number of compounds which can be represented by 1, 2, 3 and 4 ions chosen at random from 750 mass positions, if each intensity can be given a value from 1 to 10.

Table 6.4 does not, however, reflect a real-life situation. Ions are not spread randomly amongst all possible mass positions up to m/e 750, but instead are bunched at the low mass end of the spectrum. Large numbers of compounds have many ions in common in their spectra; for example, $[CH_3]^+$, $[CH_3CO]^+$, $[C_7H_7]^+$, etc. As well as this unequal spread of fragment ions, molecular weights also are not randomly distributed. The molecular weights of all the compounds listed in a yearly index of the *Mass Spectrometry Bulletin* are plotted in Fig. 6.6(a and b). Figure 6.6(a) gives the distribution of compounds with even molecular weights and Fig. 6.6(b) that for odd molecular weight compounds. For both cases, the frequency of occurrence is highest between molecular weights of 200 and 350. It is interesting that the frequency of occurrence of odd molecular weights averages out at about a quarter of that of even molecular weights, reflecting the fact that the former have to contain odd numbers of even atomic weight–odd valency or odd atomic weight–even valency atoms, which restricts the possibilities.

If fragment ions derived from these distributions of molecular weights are considered, then, since fragment ions are of necessity lower in mass than the molecular weight, the distribution curve for these will have a maximum at lower m/e than 200–350. A guess might be attempted at the maximum occurring at between m/e 12 and 100, from the usual appearance of a mass spectrum. It is a rather more complex matter to decide whether the maximum is more intense for even m/e or odd m/e fragment ions. As a general rule, it is considered that even

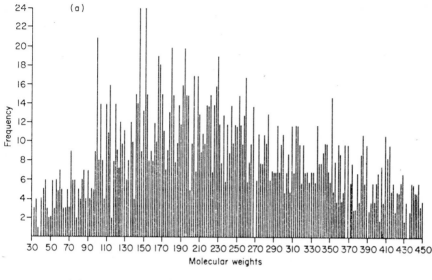

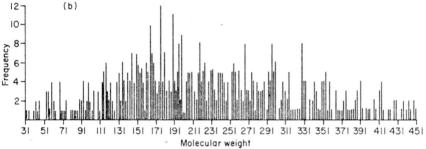

Fig. 6.6.   (a) The distribution of even molecular weights up to 450. Derived from the listing of compounds in a yearly index of the *Mass Spectrometry Bulletin*. (b) The distribution of add molecular weights up to 451, from the same source.

electron ions are more stable than odd electron ions, and it can be said that if such an ion does not contain odd atomic weight–even valency or even atomic weight–odd valency atoms, it will have an odd m/e value.

As a result of this non-random distribution of ions in a real spectrum, the figures given in Table 6.4 must be modified. To improve the uniqueness of the compound of interest compared with all the other possibilities, ions whose masses are in excess of, say, 350 should be chosen. Since the frequency of occurrence of ions of odd m/e values is a few times as high as that for even m/e values, the ion chosen should also have an even mass. If the compound has no such ions in its spectrum, it is usually possible to make a derivative which will increase the molecular weight above 350.

If these factors are taken into account, it can be said that three carefully chosen ions should distinguish the compound of interest from perhaps a million compounds chosen at random from all known organic compounds. For two carefully chosen ions, the compound would be distinguished from perhaps 10 000 compounds picked at random.

At this point it should be emphasized that the discussion has concerned the chances that the identification of a compound is correct when a limited number of ions have been chosen from its spectrum, and an unknown has been found with the same ions present with the same relative intensities. In a g.c.m.s. system, the situation will never be as clear-cut as this. On the one hand, there is a further dimension to the identification in that the compound must have the correct retention time on the particular column and under the particular conditions of temperature and flow rate. On the other hand, if identification is going to be based on intensity ratios, the occurrence of some other compound at the same retention time which happens to have a contribution at one or more of the m/e values being monitored will change these ratios, so that the correct compound will not be identified. In order to overcome this problem, the significance given to the intensities of the ions being monitored should be reduced. It is permissible to do this, since the loss in specificity by reducing the significance of intensities is amply compensated for by the requirements of correct retention time. Taking as an example a g.c. run programmed from 50–300 °C at 5 °C^{-1} min^{-1}, thus covering a range of 250 °C, it takes 50 min to complete the programme. If the g.c. resolution is such that a 15 s difference in retention time can confidently be predicted, there are 200 slots into which the compound can fall, thus improving the specificity by this factor. Since an internal standard must be used for the quantification by selected ion monitoring or retrieval, the intensity of only one of the ions being monitored from the compound itself is required in order to construct a calibration curve. If it is suspected that the intensities of some of the ions being monitored from the compound are incorrect due to a contribution from interfering substances, it is easy to determine which of the ions has the lowest interference, since it is known what the ratio of the intensities should be from inspection of its mass spectrum. This can be found by making each ion in turn have the correct intensity. Interference is lowest at this particular m/e value if the other ions have either the correct intensity or intensities which are too high. On the other hand, if there is interference at this m/e value some or all of the other ions will have intensities which appear to be lower than they should. This is illustrated for a three ion situation in Table 6.5.

TABLE 6.5
Correction of intensities of three ions in order to determine which has the lowest interference

	Required intensity (determined in an authentic spectrum)	Intensities found during selected ion monitoring		
		1	2	3
m/e a	100	100	90.48	110
m/e b	38	42	38	46.8
m/e c	22	20	18.1	22

The intensities as determined from a mass spectrum of pure compound are shown in the first column, while column 1 on the right-hand side of the table shows the intensities found by selected ion monitoring. Columns 2 and 3 are the same intensities relative to each other but multiplied by the appropriate factor to give the correct intensity for each of the ions b and c in turn. In column 1, where ion a is given the correct intensity, it is seen that ion c is too low. Therefore, there must be more interference at ion a than at ion c. In column 2, where ion b is given the correct intensity, the intensities of the other ions are too low. Therefore, there is more interference at ion b than at the other two ions. In column 3, where c is given the correct intensity, the intensities of the other two ions are both too high. This means that interference is less at ion c than at the other ions.

The result of this analysis is to show that of the three ions being monitored, c has either no interference or less than the other two ions. Thus, in calculations for calibration curves, etc. the intensity of ion c would be used. If interference at two of the m/e values is entirely absent, or identical but less than that at the other m/e value, the analysis would at some point give two correct intensities and one that was too high. Either of the two correct intensities could then be used, the more intense one having the advantage of giving an improved sensitivity for the analysis.

For the majority of cases, three carefully chosen ions will suffice for a specific determination, the intensity of that ion which is shown to have the least interference then being used for subsequent calculations. For the actual analysis, it is necessary to employ an internal standard, as discussed later (p. 135). Therefore, the same procedure should be employed on the internal standard in order to determine which ion or ions suffer the least interference. Once the optimum ion has been found for the compound to be measured and the optimum ion for the internal standard, the selected ion monitoring or retrieval needs either one channel or two channels, depending upon whether or not an internal standard is used which yields the same or a different ion from the compound to be measured. The only exception to this rule would be when large differences are expected in the amount of interference from one analysis to the next. In such a case, it is necessary to monitor as many ions as possible without serious loss of sensitivity and precision, and each determination would require a calculation as outlined above to decide which of the various ions have the least interference.

Single ion monitoring

In the discussion on sensitivity, it was pointed out that the highest sensitivity and precision was obtained with single ion monitoring, rather than with one, two, three or four channels. Many workers argue that single ion monitoring is more likely to run into trouble with interference than when several channels are used. In the previous section, however, it has been seen that by a preliminary experiment using several channels, it is possible to find an ion which suffers the least interference of all those ions which are candidates from a sensitivity point of view, i.e. the intense ions in the spectrum. If the level of compound in the samples is such that the precision is unacceptably poor when using several channels, the results of the analysis to find the best ion would be imprecise anyway. In such a rare case, single ion monitoring will be the only answer from a sensitivity point of view, and it just has to be accepted that there may be some interference at the particular m/e value being monitored.

There are several ways in which the problem of interference can be surmounted, while retaining the sensitivity advantage of single ion monitoring. For example, in the field of drug metabolism the ion or ions for analysis are frequently chosen by reference to the mass spectrum of the particular compound being studied. A more reasonable approach is to obtain a sample of urine or plasma known not to contain the drug or metabolite, and carry out the same extraction procedure on this as for the drug or metabolites. The extract is derivatized in exactly the same way, and a concentrated sample injected into the g.c.m.s. system while repetitively scanning the mass spectrum. In fact it is not strictly necessary to scan repetitively, since only a few spectra are required at the retention time of the compounds of interest. A much better judgement of which ions should be monitored can then be made by balancing the sensitivity of the analysis, which depends upon the selection of intense ions, and the occurrence of interfering ions at these values in the spectra obtained from the blank urine or plasma extracts.

This method fails where some endogenous compound is to be measured, so that a blank sample which does not contain the substance cannot be obtained. In such a position, low resolution mass spectrometry will not determine how much interference there is at a particular m/e value characteristic of the compound to be measured. A laborious method of choosing the ion at which the interference was least would be to employ the calculation method for the three ion situation, where the m/e value at which interference was minimal could be determined. It may be necessary to examine 20 or 30 ions in this way, relating them to the intensities obtained previously on a pure sample of the compound, before an acceptable ion can be found for the analysis.

A more reliable method for decreasing the contribution made at the measured m/e value by interfering substances is to use higher mass spectrometric resolving power if such an instrument is available.

The definition of resolving power $R = M/\Delta M$, where M is the mean position of two adjacent peaks, and ΔM the separation between them is usually quoted for a 10 % valley between the peaks (Chapter 1). Thus, at a resolving power of 1000 and integral m/e value of 200, two ions of m/e 200.00 and 200.2 are separated by a 10 % valley. If the peaks are considered to be triangular, in order to simplify the picture, it can be seen for such a pair of peaks shown in Fig. 6.7(a) that when the instrument is tuned to the top of one peak, the contribution made by the other peak is zero. For a Gaussian shaped peak, the contribution would not be zero, but nevertheless extremely small.

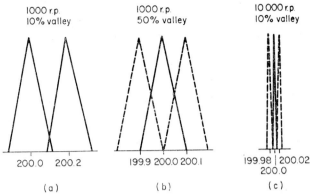

Fig. 6.7.   The contribution made by an adjacent peak when tuned to the maximum of a peak at m/e 200.00. It is assumed that the peaks are triangular for the sake of simplicity. (a) At 1000 resolution, there is no contribution by the adjacent peak of m/e 200.2 to the maximum, even though the overlap amounts to a 10% valley. (b) At 1000 resolution, an adjacent peak with an overlap amounting to a 50% valley just starts to contribute to the maximum. (c) At 10 000 resolution, there is no contribution by an adjacent peak of m/e 200.02 which overlaps with a 10% valley.

The situation can now be considered where the valley between adjacent peaks is 50% before the cross-contribution becomes significant at the top of either peak. Under the conditions of 1000 resolving power, there is a mass window of 0.2 mass units outside of which other ions make a negligible contribution. This window is inversely proportional to the resolving power of the instrument, as shown in Table 6.6. The effect of operating at 10 000 resolving power rather

TABLE 6.6
Width of the mass window covered when various m/e values are monitored at different resolving powers

Resolving power	Mass window at m/e				
	100	200	300	400	500
1000	0.1	0.2	0.3	0.4	0.5
1500	0.067	0.13	0.2	0.27	0.33
3000	0.033	0.067	0.1	0.133	0.17
7500	0.013	0.026	0.04	0.053	0.067
10000	0.01	0.02	0.03	0.04	0.05
20000	0.005	0.01	0.015	0.02	0.025

than 1000 is to decrease the mass window by a factor of 10. This in turn decreases the chances by a factor of 10 that any other compound will have an ion with an m/e value sufficient to place it within the mass window being monitored. The improvement in specificity gained by increasing the resolving power from 1000 to 5000 can be seen in Fig. 6.8, where an ion at m/e 174.1 is monitored from a urine extract.

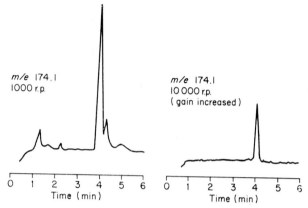

m/e 174.1
1000 r.p.

m/e 174.1
10 000 r.p.
(gain increased)

Time (min)   Time (min)

Fig. 6.8.   The increase in specificity obtained when monitoring an ion at m/e 174.1 in a
urine extract by increasing the resolution from 1000 to 10 000.

This technique has been used for quantifying C_{19} steroids in hyperplastic
prostate tissue.[13] Figure 6.9 shows the selected ion profiles obtained from a
standard mixture of 5α-androstanediols and from a prostatic tissue extract by
monitoring m/e 436.319 formed by their TMS derivatives. Under these conditions, the peak heights in the standard mixture were reproducible to within an

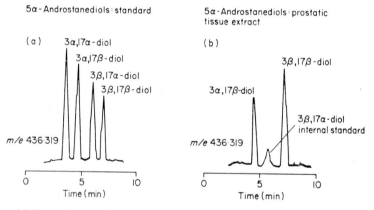

5α-Androstanediols: standard

(a)   3α,17α-diol
       3α,17β-diol
       3β,17α-diol
       3β,17β-diol

m/e 436.319

Time (min)

5α-Androstanediols: prostatic
tissue extract

(b)   3β,17β-diol

3α,17β-diol

3β,17α-diol
internal standard

m/e 436.319

Time (min)

Fig. 6.9.   (a) Separation and detection of the isomeric 5α-androstanediols (0.2 ng each
approximately) as their TMS derivatives, by g.c. high resolution single ion monitoring.
(b) Detection of 5α-androstane-3α,17β-diol and 5α-androstane-3β,17β-diol in a typical
extract from benign hyperplastic prostate tissue to which 2 ng of internal standard
(5α-androstane-3β,17α-diol) was added before extraction. (Reproduced with permission
from Ref. 13.)

accuracy of $\pm 5\%$ on 200 pg quantities of each component. Twenty pg of
5α-androstane-3β,17β-diol was detectable with an S/N ratio of better than 3:1.
This reduction of the interference from other ions of the same nominal mass

value must be offset against the decreased sensitivity obtained for the analysis at higher resolving powers. The sensitivity is not directly proportional to the resolving power, but follows a curve such as that shown in Fig. 6.10 for an MS-30 instrument.

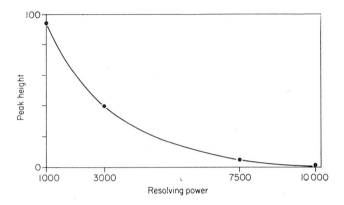

Fig. 6.10. Response obtained at various resolving powers at m/e 174.1 ($[CH_2=N(TMS)_2]^+$) for 20 pg injections of octopamine as the tetrakis-TMS derivative.

The curve was obtained by injecting 20 pg amounts of the tetra-TMS derivative of octopamine into the g.c.m.s. The ion at m/e 174 due to $[CH_2=N(TMS)_2]^+$ was monitored at resolving powers of 1000, 3000, 7500 and 10 000 without changing the gain on the multiplier or pen recorder. For the 20 pg injection at 10 000 resolving power, the S/N ratio was better than 5:1.

Compared with the sensitivity at a resolving power of 1000, that at 10 000 is extremely low. However, as can be seen from the curve in Fig. 6.10, the sensitivity at 5000 resolving power is down only by a factor of four, making operation in the range of 3–5000 an attractive proposition for the reduction of interference from ions of the same nominal mass.

Internal standards, their advantages and disadvantages

Type A standards, stable isotope labelled analogues

Since a stable isotope labelled analogue of the compound to be measured has such similar physical and chemical properties to the unlabelled compound, it would seem to be ideal as an internal standard, for the following reasons:

(a) *The partition coefficients are similar.* Although not much evidence to this end is presented in the literature, for the majority of compounds the partition coefficients for labelled and unlabelled compound are thought to be the same. This removes a possible source of error arising through inefficient and variable extraction procedures. Since the partition coefficients can be taken as identical, the ratio of compound to internal standard will remain constant even though the extraction conditions may be changed inadvertently from one run to the next. For example, 4 ml of solvent may be used instead of 5 ml, or the mixture shaken for 1 min instead of 5, but such differences will have little effect on the internal standard/compound ratio.

(b) *The rate of derivatization is similar.* Except for those cases where the cleavage of a bond to a labelled atom is involved, or where a secondary isotope effect is rather large, unlabelled and labelled analogues should form derivatives at the same rate. Thus, a change in reaction temperature, or solvent, or length of time for which the reaction is allowed to proceed has a minimum effect upon the ratio of unlabelled and labelled derivatives as compared with the ratio of the original compounds.

(c) *The retention times on the g.c. column are similar.* The closer the retention times of compound and its internal standard, the slighter will be the effect of variations in instrumental parameters. The changes which can occur in these parameters have been discussed in Chapter 4, but for the most part they tend to occur over minutes or hours rather than seconds. The variation in the responses obtained for two peaks from such causes are likely to be much greater if the peaks are separated by 5 min than if they elute within seconds of each other. Since the unlabelled and labelled compounds elute closely together, the effect of instrumental changes is at a minimum.

It is difficult to predict the difference in retention time between a compound and its labelled analogue, since some analogues with only one deuterium atom are quite obviously resolved from the unlabelled compound, while others with a large number of deuteriums co-elute with unlabelled compound on a variety of columns.

(d) *Carrier effect.* The carrier effect can be represented in diagrammatic form as shown in Fig. 6.11(a and b).

If smaller and smaller amounts of a substance are extracted from solution, perhaps introducing a derivatization step, and then single ion monitoring performed on the total extract, using an intense ion in its spectrum, the responses resemble those in Fig. 6.11(a). Initially, for the larger amounts extracted, the responses decrease linearly, but reach a point at which they tail off rapidly, with a lower limit of detectability of perhaps 100 pg in the original solution.

A stable isotope labelled analogue can then be chosen, labelled in such a way that the diagnostic ion is shifted from its original value and, say, 1000 times the amount of this added to each solution that is extracted. If there is a carrier effect and the procedure is repeated as before, monitoring the ion formed by the unlabelled compound present, it will be seen that the linear portion of the responses, as shown in Fig. 6.11(b), now extends to much lower levels of unlabelled compound. In addition to this, it can be seen that a much lower level, down to a few picograms, can be detected.

The effect of the labelled compound is therefore that small amounts of the unlabelled compound, which would otherwise be lost, are carried through the whole extraction, derivatization and g.c.m.s. procedure.

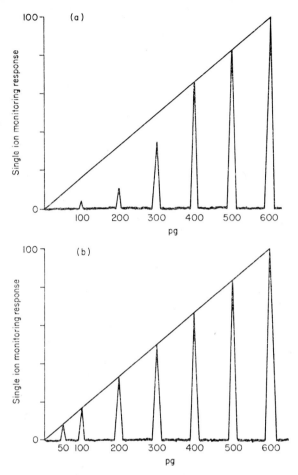

Fig. 6.11.   (a) Response at a particular *m/e* value for successively smaller injections of a compound into a g.c.m.s. system. (b) Response at the same *m/e* value for the same injections of the compound in the presence of a large excess of a labelled analogue, assuming a carrier effect to operate.

Suppose a small amount, *w*, of substance is subjected to the total procedure. If the absorption capacity of the whole system is *c*, then the amount *a* entering the source of the mass spectrometer is given by

$$a = w - c$$

As long as $w > c$, a response will be seen, but naturally, once $w = c$, the response vanishes.

If an amount of carrier *n* times the amount of substance *w* is added, the definition of a carrier being a substance which the absorbing system cannot distinguish from the original compound (although the mass spectrometer can by virtue of a different mass spectrum), then the amount of original substance absorbed is now $c/(n + 1)$ and the amount of carrier absorbed is $nc/(n + 1)$.

Thus, the amount of original substance entering the mass spectrometer is given by

$$a = w - \frac{c}{(n + 1)} \approx w \text{ if } n \text{ is large}$$

Typical values for n, the amount of carrier added, lie between 100 and 1000.

Because of these advantages, the use of stable isotope labelled analogues, especially for the determination of drugs and metabolites in biological fluids, has mushroomed over the last few years. Levels of a few nanograms per ml of plasma of such compounds have been readily determined, the lowest concentrations being 80 pg ml^{-1} of plasma for phenobarbital and valium[14] and 50 pg ml^{-1} for scopolamine.[15] Some typical determinations using type A standards are given in Table 6.7.

TABLE 6.7
Some typical determinations of drugs and other biologically important compounds using type A (stable isotope labelled) internal standards

Compound	Isotope	Levels measured	Reference
Nortriptyline	^{2}H, ^{15}N	5 ng ml^{-1} plasma	16
Amphetamine	^{2}H	2–5 pmol ml^{-1} plasma	17
Phentermine	^{2}H	3–8 pmol ml^{-1} plasma	17
Diphenylhydantoin	^{13}C	11 μg ml^{-1} plasma	14
Pentobarbital	^{13}C	80 pg ml^{-1} plasma	14
Valium	^{13}C	80 pg ml^{-1} plasma	14
Amitriptyline	^{2}H	10 ng ml^{-1} plasma	18
Nortriptyline	^{2}H	10 ng ml^{-1} plasma	18
Imipramine	^{2}H	10 ng ml^{-1} plasma	18
Morphine	^{2}H	1 ng ml^{-1} plasma	19
Phosphoramide mustard	^{2}H	20 ng ml^{-1} serum	20
Δ^9-THC	^{2}H	300 pg ml^{-1} plasma	21
5-Hydroxyindole-5-acetic acid	^{2}H	2 ng ml^{-1} CSF	22
Prostaglandin F$_{2\alpha}$	^{2}H	400 pg injection	23
Prostaglandin F$_{1\alpha}$ and prostaglandin F$_{2\alpha}$ major metabolite	^{2}H	510 ng injection	24
Biogenic amine metabolites	^{2}H	0.4–154 ng ml^{-1} CSF	25

Most determinations employing type A standards use an analogue of the original compound, and these may be quite expensive to synthesize. There are a few examples of the use of labelled derivatives,[21,26] which are much more economical to prepare. It is interesting that the first published paper on the use of isotopically labelled carriers involved the use of a labelled derivatizing agent.[26] The O-methyloxime methyl ester of prostaglandin E$_1$ (12) was added to an amount of O-[^{2}H$_3$]methyloxime (13) and the mixture then trimethylsilylated before selected ion monitoring of the ions at m/e 470 and 473 formed by loss of C$_5$H$_{11}$· from the molecular ions. The relationship between peak areas and amount of unlabelled prostaglandin E_1 from which the derivative was made was found to be linear in the range of 3–10 ng injected on the column. This experiment concerned the injection of pure mixtures on the column, and no extraction from biological fluids was attempted. There appeared to be a carrier effect demonstrated purely for passage through the g.c.m.s. in this work. It is logical that there is also a carrier effect for extraction and derivatization, so a complete determination of a compound in a biological fluid may involve two distinct carrier effects.

It has been shown recently that it is not necessary to have a carrier to pass small amounts of some compounds through the g.c.m.s. system,[9,27,28] and although only a limited number of compounds and columns have been studied, the technique may well be of more general applicability. The attainment of high sensitivity without the use of carriers is discussed later (p. 152).

Since carriers are not necessary for some compounds, this raises the question of whether there is such a thing as a carrier effect. Although such an effect has

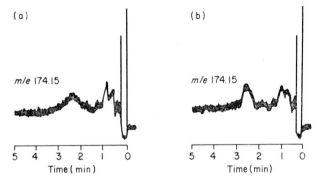

Fig. 6.12. (a) The response at m/e 174·1 for an injection of 20 pg of the tetrakis-TMS derivative of octopamine on a newly conditioned column. (b) the response at m/e 174.1 for 20 pg of the tetrakis-TMS derivative of octopamine in the presence of 2 ng of the same derivative of [$^2H_3$]octopamine.

been assumed by many workers, since a large number of papers describe the use
of 100–1000 times the amount of labelled analogue to act as a carrier, careful
scrutiny of the literature shows that a carrier effect has never been demonstrated.
An experiment was carried out in these laboratories to determine whether such
an effect existed for octopamine and $[^2H_3]$octopamine, analysed as the tetrakis-
TMS derivatives (**14** and **15**), and it was demonstrated that no carrier effect
operated when 100 times the excess of labelled analogue was used.[28]

The operation of a carrier effect through a g.c. column is best demonstrated

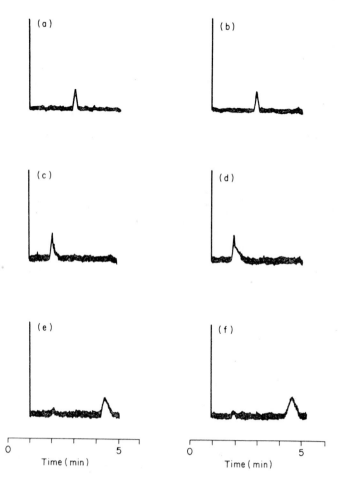

Fig. 6.13. Absence of a carrier effect: (a) m/e 423 from 20 pg of prostaglandin $F_{2\alpha}$
methyl ester TMS derivative on a 1% SE 30 column; (b) the same injection in the
presence of 2 ng of the $[^2H_4]$ analogue; (c) the molecular ion at m/e 318 from 1 pg of
methyl arachidonate on a 1% OV 101 column; (d) the same injection in the presence
of 1 ng of the $[^2H_8]$ analogue; (e) m/e 295 from 50 pg of prostaglandin E_2 methyl ester
methyl oxime TMS derivative on a 1% OV 17 column; (f) the same injection in the
presence of 5 ng of the $[^2H_4]$ analogue.

for levels of compound which are only just detectable in the absence of carrier. Mass spectrometers are generally tuned for the particular ion used in the analysis by injecting fairly large amounts of the compound. This, however, then improves the sensitivity of the column towards that compound, so that it is necessary to tune in with such injections on a different column. A few small injections of **14** showed that the detection limit was about 20 pg on a freshly prepared and conditioned column. The response for this injection is shown in Fig. 6.12(a).

An injection of 20 pg of **14** plus 2 ng of **15** was then carried out, giving the response shown in Fig. 6.12(b). Since the **15** injected contained 10 pg of **14** as an isotopic impurity, the injection was equivalent to 30 pg of **14**. The response in Fig. 6.12(b) is only slightly more than 1.5 times the response in Fig. 6.12(a), so the carrier effect was insignificant.

Following the results with octopamine, the effect was tested with a number of other compounds and their labelled analogues, some of which have been described in the literature as carriers. In no case was a carrier effect observed when the signal from a barely detectable level of the unlabelled compound was compared with the signal obtained in the presence of amounts from 100 to 1000 times the labelled analogue. Some representative examples are shown in Fig. 6.13.

There is no difference between the signal at m/e 423 for 10 pg of prostaglandin $F_{2\alpha}$ methyl ester tris-TMS derivative on its own and in the presence of 2 ng of the $[^2H_4]$ labelled compound. The same is true for the molecular ion at m/e 318 from 1 pg of methyl arachidonate on its own and with 200 pg of the $[^2H_8]$ analogue, and for the fragment ion at m/e 295 from 50 pg of prostaglandin E_2 methyl ester methyl oxime TMS derivative on its own and in the presence of 5 ng of the $[^2H_4]$ analogue.

These results raise the question of whether there is a carrier effect with deuterium labelled compounds. A carrier has been defined previously as a compound which the absorbing system is unable to distinguish from the compound of interest. However, most deuterium labelled compounds exhibit a small separation from the unlabelled compound on the g.c. column. This is sometimes difficult to see on the total ion monitor, but can be observed if an ion from the unlabelled compound and an ion unique to the labelled compound are observed on separate channels. Thus, the column can distinguish the deuterium labelled and unlabelled compounds, so perhaps it is not surprising that in the above examples the deuterium labelled compounds do not act as carriers.

The situation may well be different for ^{13}C labelled compounds, but the problem with these is the relatively low isotopic purity of the starting materials, usually about 60%. It will be necessary to incorporate several atoms before amounts such as 100 to 1000 in excess of the labelled compound can be used to check for a carrier effect.

When a carrier effect can be demonstrated, so that many times the amount of labelled compound is used, it is an advantage to be able to adjust the dwell times on each peak, so that more time is spent on the weaker compound peak and less time on the intense internal standard peak. This can lead to an improvement in precision, and is shown by an example worked out by assuming that low numbers of ions are counted during a g.c. peak width of up to 30 s.[29]

Let the compound have a diagnostic ion at m/e m_1, and let the isotopic dilution compound have a diagnostic ion at m/e m_2, the concentration of the labelled compound being about 10 times that of the unlabelled compound. Let $\lambda = 540$ total counts (net) during a 5 s total transit time at m/e m_2, and $k\lambda = 54$ total counts (net) during a 5 s total transit time at m/e m_1. Therefore, $k = 0.10$.

The fractional dwell time at $m_2 = U_1 = \dfrac{\sqrt{k}}{1 + \sqrt{k}} = 0.24$

and

$$U_2 = 1 - U_1 = 0.76$$

The improvement in precision obtained by using these dwell times U_1 and U_2 rather than equal dwell times, as is more usual, is

$$\frac{2(1 + k)}{(1 + \sqrt{k})^2} = 1.27$$

Table 6.8 shows the result of similar calculations for other dilution ratios.

There are three major disadvantages in using stable isotope labelled internal standards; first their cost, second the cost of the equipment necessary to switch

TABLE 6.8
Optimum dwell times on each of two channels and precision improvement factor[29]

Sample/isotope standard	k	Fractional dwell time		Precision improvement factor
		Sample channel	Isotope channel	
1:100	0.01	0.91	0.09	1.67
1:10	0.1	0.76	0.24	1.27
1:1	1.0	0.50	0.50	1.00
2:1	2.0	0.42	0.58	1.03
10:1	10.0	0.24	0.76	1.27

the mass spectrometer, and third the poorer precision in the g.c.m.s. determination compared with single ion monitoring.

(a) *Cost of labelled compounds.* Although the simpler starting materials such as $LiAl^2H_4$, $CH_3{}^{13}CO_2H$, $^{13}CH_3CO_2H$, CH_3O^2H, C^2H_3I, etc., are relatively cheap, most labelled standards require many stages for their synthesis, and their cost can be measured in hundreds of dollars for a few hundred milligrams. As pointed out earlier, deuterium labelled standards, although cheaper than ^{13}C labelled compounds, may not exhibit the highly desirable carrier effect, so that nothing is gained by their use in preference to type B or C standards. ^{13}C compounds, which probably do act as carriers, become extremely expensive when a high degree of isotopic purity is required. The amount of unlabelled material present must be minimal if linear calibration lines are to be achieved. Since ^{13}C labelled starting materials will not be as isotopically pure as ^2H compounds, then in order to reduce the amount of unlabelled material present it may be necessary to incorporate three or more ^{13}C atoms. This may turn out to be an extremely difficult process, necessitating a totally different synthetic route from that employed for the unlabelled compound, and the cost may be unacceptably high. It is much easier to incorporate multiple deuterium atoms, which can be introduced by way of trideuteromethyl groups or by exchange of aromatic ring hydrogens, for example.

A more economical approach to the preparation of isotopically labelled analogues is to employ labelled derivatizing agents. Thus, $[^2H_3]$methyloximes are readily prepared from ketones, $[^2H_2]$diazomethane can be used for methylation, and $[^2H_9]$trimethylsilylating agents for the preparation of labelled TMS ethers and esters. One disadvantage of this approach, however, is that such derivatives cannot be added readily at the earliest point in an analytical procedure, but are usually introduced at the point at which derivatization is carried out. By doing this, the variation in losses at the prior extraction stages is not compensated for, since no internal standard is present at that point.

(b) *Cost of accelerating voltage alternators or other switching devices.* With a labelled internal standard which has a diagnostic ion a few mass units away from the ion produced by the compound of interest, it is necessary to have a device to switch the mass spectrometer between these two masses. Most quadrupole mass spectrometers are delivered with such a capability, but accelerating voltage alternators for magnetic instruments may cost around $15 000 for six channels.

(c) *Poorer precision.* Besides the disadvantages of using two channels rather than one at low compound levels discussed earlier (p. 126), there is also the point that drift on each channel may occur at a different rate, and even in opposite directions, due to changes in the circuit characteristics of the switching equipment. This will, of course, be superimposed upon the normal magnet drift of magnetic instruments. Over the course of several hours this would mean a change in the ratio of the two responses for the same solutions injected. A similar error is also introduced if one channel is tuned correctly and one is mistuned, since later determinations when the two channels are correctly tuned will give a different response ratio. This is not a disadvantage in single ion monitoring since drift at one m/e value should be negligible over a period of minutes. As pointed out in Chapters 3 and 4, quite frequently the mass spectrometer precision is of less importance than the poorer precision in other stages of the analysis.

Type B standards, homologues yielding a common ion

Next to stable isotope labelled internal standards in terms of popularity are homologues which can yield an ion of the same m/e value as the compound of interest. These have the great advantage that their use requires no switching equipment, so that any mass spectrometer can be used. Synthetically they pose no problems, and can be made at a fraction of the cost of stable isotope labelled analogues. Usually the synthetic route to the compound of interest can be joined only a few stages from the end-product to introduce, for example, diethylamino groups in place of dimethylamino, or *p*-tolyl instead of benzyl. The groups must be placed so that they do not shift the m/e value of the ion of interest.

Partition coefficients. Usually, a homologue which has a retention time on g.c. fairly close to the original compound will have similar extraction properties. As an example, mixtures of mepyramine (16) and its dibutyl analogue (17) are extractable to the same extent from plasma. Table 6.9 shows the ratio of the

TABLE 6.9
Extraction of mepyramine (16) and its dibutyl analogue (17) from plasma by ethyl acetate. Responses at m/e 121 from each compound used to calculate ratios in extract

Ratio of 16:17 in spiking mixture	Ratio of 16:17 in extract
10.0	9.95
3.0	3.01
1.0	1.0
0.33	0.34
0.1	0.1

two compounds in the spiking mixture added to plasma, and the ratio of these two in the ethyl acetate extract, determined by single ion monitoring. The ratios remain the same within the limits of experimental error.

16   17

If the partition coefficients are quite different, great care will have to be taken to ensure that the extraction conditions are reproduced exactly. It is necessary to make sure that the calibration curve is constructed by extracting known amounts of compound and standard added to urine or plasma, etc. and determining the ratio of the two components in the extract, and that calibration extends both above and below the concentrations likely to be encountered with unknowns.

Gas chromatographic retention times. The effect of drift from the top of the mass spectral peak being monitored is at a minimum when the compound and its internal standard elute within a short time from the g.c. column. A reasonable

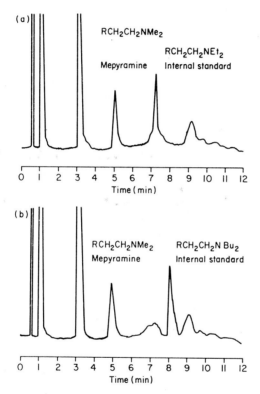

Fig. 6.14.   Single ion monitoring of m/e 121 from a plasma extract of a patient taking mepyramine: (a) with a diethyl homologue as internal standard, there is an interfering peak superimposed on the internal standard response; (b) with a di-*n*-butyl homologue as internal standard, the internal standard is at a part of the chromatogram free of interference.

difference in retention times of the two components would be that necessary to achieve complete resolution plus another minute. No advantage is gained by having a wider separation than this unless some interfering peak is present at the retention time of the first choice of internal standard. A higher homologue may then be used to move the internal standard to a clearer part of the chromatogram. This was necessary in the case of mepyramine in plasma. Single ion monitoring of m/e 121 derived from mepyramine and its internal standard, the diethyl homologue, revealed that an interfering peak was present at the same point as the signal from the homologue (Fig. 6.14(a)). The problem was overcome by using the dibutyl homologue (17) (Fig. 6.14(b)).

Type A and B standards have been compared for the quantification of the barbiturate allobarbitone on a quadrupole mass spectrometer using the dimethyl derivative (18).[30,31] The [^{2}H$_6$]dimethyl derivative (19) was used as a type A standard and dimethylquinalbarbitone (20) as a type B standard. Since, at that time, it was thought that a carrier was necessary for low levels of 18, the comparison involved injection of mixtures of 18, 19 and 20. In the first case 19

18   19   20

was both internal standard and carrier, while in the second case 20 was the internal standard and 19 the carrier. The first experiment monitored the two ions m/e 195 and 199 (m/e 199 was the most intense ion in isotopically impure 19) formed by 18 and 19, and the results were compared with those from a second experiment in which single ion monitoring of m/e 195 from both 18 and 20 was carried out. The results are shown in Table 6.10.

TABLE 6.10
Coefficients of variation of the determination of dimethylallobarbitone (18) by co-injection with deuteriated dimethylallobarbitone (19) and dimethylquinalbarbitone (20). Five injections of each solution

Amount of 18 injected (ng)	0.04	0.1	0.4	1.0
Selected ion monitoring with 19 as standard[a]	19.8	15.3	1.9	1.3
Single ion monitoring with 20 as standard	11.5	2.0	0.8	0.6

[a] A four-channel device was used with two channels duplicated.

At all points the coefficients of variation for single ion monitoring are superior to those for four-channel monitoring with two channels duplicated, and the difference is particularly significant at the 100 pg injection of 18. In this work, only slightly less than 40 pg could be detected when in the four-channel switching mode but down to 10 pg could be detected by single ion monitoring.

These experiments show that a homologous internal standard, with which single ion monitoring can be employed, is superior to labelled standards on grounds of sensitivity and precision of the mass spectrometric determination. One major reason for this is the effect of ion statistics at low sample levels. Other reasons are the instability introduced by switching the instrument between channels, and the fact that the drift on each channel may not be synchronized.

Although the effect of ion statistics is unavoidable, superior instrumentation can reduce instability and drift, so that the difference between single and multiple ion monitoring may not be so marked as in the allobarbitone case, which was carried out on an older instrument. It is difficult to predict the difference between the two for extractions from biological fluids, since much will depend upon the physicochemical properties of the various compounds, and the presence or absence of interfering substances.

Some typical applications of single ion monitoring using type B standards are given in Table 6.11.

TABLE 6.11
Some typical applications of type B (homologues yielding a common ion) internal standards

Compound	Internal standard	Levels measured	Reference
Amylobarbitone	Butobarbitone	10 ng ml^{-1} plasma	32
Lidocaine	Trimecaine	0.5 μg ml^{-1} plasma	33
Desethyl lidocaine	Lidocaine	0.3 μg ml^{-1} plasma	33
Butobarbitone and metabolites	Pentobarbitone	—	34
Pentazocine	Cyclazocine	2 ng ml^{-1} CSF	35
Propanolol	Oxprenolol	1 ng ml^{-1} plasma	36
Prostaglandin F$_{2\alpha}$	ω-Trinor-16-cyclo hexyl prostaglandin F$_{2\alpha}$	2 pg injection	27
Prostaglandin F$_{2\alpha}$	ω-Trinor-16-cyclo hexyl prostaglandin F$_{2\alpha}$	50 fg injection	9
Testosterone[a]	Epitestosterone	1.7 ng per g of prostate tissue	13

[a] At 10 000 resolving power.

Type C internal standards, compounds yielding a different ion from the compound of interest

The use of a type C internal standard, which is a compound having similar extraction and g.c. properties to the compound of interest but yielding a different ion, would appear to have little to recommend it. In most cases, its use means that two channels are needed for the measurement, so that the cost and other advantages of single ion monitoring are not achieved. Furthermore, while differential drift on two channels can be of some concern with type A standards, which elute almost simultaneously, its effect on two peaks which elute perhaps several minutes apart, as with type C standards, is certain to be much more serious.

Unless the type C standard is a homologue with a substituent placed so as to shift the diagnostic ion, it is unlikely to have a partition coefficient as close to the compound of interest as it typical for homologues, although, as pointed out earlier, this is not a serious disadvantage if care is taken over calibration curves. One advantage of such a standard is that it is readily chosen from the variety of compounds available in most laboratories, so that no synthetic work is necessary. Some typical applications of type C standards are given in Table 6.12.

TABLE 6.12
Some typical applications of type C standards (compounds yielding a different ion)

Compound	Internal standard	Levels measured	Reference
Hydroxyamylobarbitone	Phenobarbitone	50 ng ml^{-1} plasma	32
Indole-3-acetic acid	5-Methylindole-3-acetic acid	2 ng ml^{-1} CSF	37
Imipramine	Promazine	10 ng ml^{-1} plasma	38
Piribedil	1-Benzylamino-4-chloro-benzophenone	10 ng ml^{-1} plasma	39
Methadone	2-Dimethylamino-4,4-diphenyl-octan-5-one	16 pmol ml^{-1} plasma	40

It is possible to utilize the advantage of single ion monitoring for type C standards by retuning the mass spectrometer to the new m/e value once the first peak, from compound or internal standard, has eluted. This approach was used

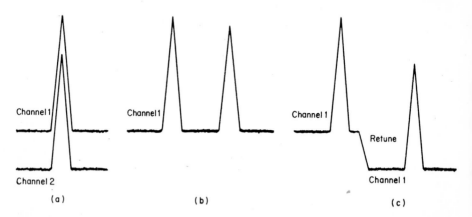

Fig. 6.15. A representation of the information obtained by the use of the three types of internal standard. (a) Type A standard, a stable isotope labelled analogue of the compound of interest yielding a different ion at approximately the same retention time. (b) Type B standard, a homologue or the compound of interest yielding an ion of the same m/e value at a different retention time. (c) Type C standard, a chemically similar compound yielding a different ion at a different retention time.

for the determination of 3'-hydroxyamylobarbitone in plasma with pheno-barbitone as internal standard.[32] The diagnostic ion for the hydroxy compound was at m/e 169, and once this peak had eluted the instrument was retuned to m/e 175 for an intense ion in the spectrum of phenobarbitone. However, this approach was subject to considerable error inherent in the retuning process. With a quadrupole instrument or a magnetic instrument fitted with an accelerating voltage alternator, it is convenient to tune in the two ions to be used for the analysis on separate channels by using injections of authentic compounds. The channel can be switched at the appropriate time between peaks. This method should lead to greatly reduced error compared with manual tuning between peaks.

In conclusion, the information obtained by using variously type A, B and C internal standards with a mass spectrometer capable of monitoring more than one channel is shown in Fig. 6.15.

Chemical ionization

In the course of the last few years, chemical ionization facilities have become available to a large number of laboratories interested in quantification. The advantages of chemical ionization over electron impact ionization for quantification are highlighted by the two spectra shown in Fig. 6.16 for 2-ethyl-2-phenyl-malondiamide, a metabolite of primidone.[41]

The molecular ion is only 0.6% of the base peak in the e.i. spectrum, but the $[M + 1]^+$ ion is the base peak in the c.i. spectrum. The ion of highest m/e value in the e.i. spectrum is at m/e 163 and accounts for about 20% of the total ion current; thus, it would confer only average sensitivity and specificity to an analysis by selected ion monitoring. The $[M + 1]^+$ ion in the c.i. spectrum carries about 70% of the total ion current due to the compound (the ion at m/e 57 is derived from the isobutane reagent gas), so that such an ion imparts very much higher sensitivity to the analysis. This ion, being at a slightly higher m/e value than the highest useful ion in the e.i. spectrum, should be slightly less likely to suffer interference from endogenous substances in biological extracts.

In any discussion of chemical ionization, the question inevitably arises as to whether it is a more sensitive technique than electron impact. A recent paper[42] on the mass spectrometry of the cyclic boronate derivatives of prostaglandin F_α compared the signals obtained with single ion monitoring under both ionization modes. Prostaglandin $F_{2\alpha}$ methyl ester methane boronate TMS ether gave an $[M - 71]^+$ ion under e.i. conditions that was about 5–10% of the total ion current, as estimated from the published spectrum. Under c.i. conditions the base peak at m/e 315, due to $[M - RBO_2H_2 - TMSO]^+$ accounted for about 40–50% of the total ion current. By comparing the peak heights obtained on the same amount of sample, it was found that the ratio of the signals in c.i./e.i. mode was 5.4:1, as shown in Fig. 6.17. It was found also that the signal response was linear with respect to the amount of sample over the range 500 pg to 1 μg.

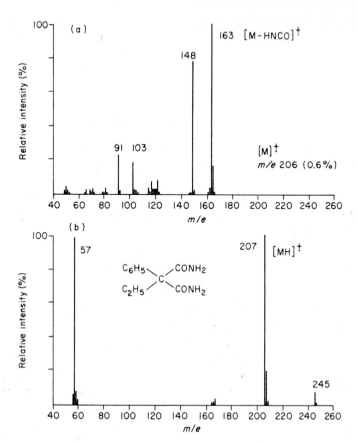

Fig. 6.16. A comparison of the e.i. (a) and isobutane c.i. (b) mass spectra of a metabolite of primidone. (Reproduced with permission from Ref. 41.)

TABLE 6.13

Some typical application of chemical ionization to quantitative determinations of drugs

Compound	Reagent gas	Internal standard	Levels measured	Reference
Phenformin	Methane	A	1 ng ml^{-1} plasma	43
Morphine	Methane	A	5 ng ml^{-1} plasma	44
Phencyclidine	Methane	A	1 ng ml^{-1} plasma	45
Imipramine	Methane	A	55 ng ml^{-1} plasma	46
Diphenoxylic acid	Methane	A	10 ng ml^{-1} plasma	47
Amphetamine	Methane	A	10 ng ml^{-1} plasma	48
Phenylbutazone	Isobutane	A	500 ng ml^{-1} plasma	49
Phosphoramide mustard	Isobutane	A	20 ng ml^{-1} serum	20

In terms of absolute numbers of ions produced in the source, it appears that electron impact has a slight edge, since the % total ion current of the peaks in question are in the ratio of 40 or 50 to 5 or 10, but this depends very much upon the design of the sources concerned. Because in a typical e.i. spectrum the base peak probably accounts for 10–20% of the total ion current, while in a typical c.i. spectrum it accounts for, say, 70–80% of the total ion current, chemical ionization can be considered to be usually the more sensitive technique.

The majority of quantitative work carried out using chemical ionization has employed type A standards, and some representative papers are referenced in Table 6.13.

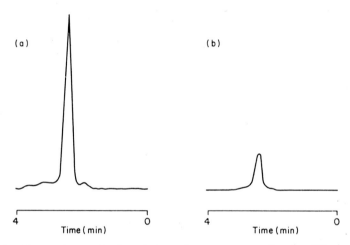

Fig. 6.17. Comparison of selected ion chromatograms (by repetitive scanning) obtained during g.c.m.s. of prostaglandin $F_{2\alpha}$ methyl ester methane boronate TMS ether: (a) m/e 315 ($[M - CH_3BO_2H_2 - TMSO]^+$) during c.i. (isobutane); (b) m/e 393 ($[M - 71]^+$) during e.i. at 70 eV. (Reproduced with permission from Ref. 42.)

Capillary columns

The use of glass capillary columns coupled to mass spectrometers is in its infancy at present, and reports are still awaited on their application to quantitative determinations. Several methods have been reported for the direct coupling of capillary columns to the mass spectrometer.[50-52] Besides the obvious advantage of superior resolution, which can reduce the contribution of interfering compounds at the position of the peak of interest, they will increase the sensitivity of the analysis compared with the same levels on a packed column by virtue of narrower peaks and hence higher momentary sample levels entering the mass spectrometer. Eyem[10] has developed the equation

$$h = \frac{A}{t_R} \sqrt{\left(\frac{N}{2\pi}\right)}$$

where h is the signal height, A is the amount of sample, t_R is the retention time of the compound, and N the efficiency of the column.

Thus, the ratio of the signal heights for the same injections on two columns is proportional to the square root of the ratio of the efficiencies. A typical packed column has approximately $N = 6000$, while 100 000 is more typical for a good capillary column. Thus, the signal height from a capillary column is about four times that from a packed column for the same amount of sample injected.

The bleed from capillary columns is much less than that from packed columns, so that an even greater advantage accrues from their use for the determination of low sample levels.

Femtogram gas chromatography mass spectrometry

Assuming that the extraction and derivatization conditions have been maximized, the attainment of high sensitivity in selected ion monitoring depends upon

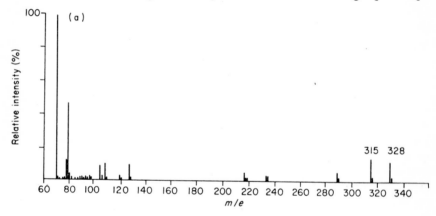

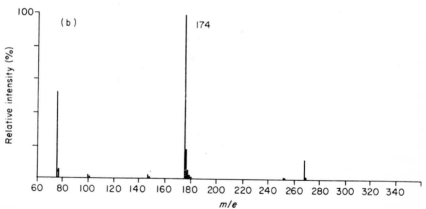

Fig. 6.18. The advantage of a careful choice of derivative for selected ion monitoring under e.i. conditions; (a) the tris-trifluoroacetyl derivative of octopamine; (b) the tetrakis-TMS derivative of octopamine.

the efficient conversion of a certain amount of compound injected onto the gas chromatograph into as many ions as possible at one particular m/e value. In order to achieve this, loss of compound on the g.c. column and at the separator should be minimal, while the mass spectrometer should be as effective as possible at converting the sample which eventually arrives at the source into a signal at the recorder.

The mass spectrometer should be in peak condition, with the source as clean as possible, and the metal surfaces likely to come into contact with sample highly polished. The electron multiplier should have a gain in excess of 10^6, while the noise level for the multiplier, amplifiers and recording system should be as low as possible. Any well maintained instrument should satisfy these criteria, so that the main problems to be overcome are the elimination of losses in the gas chromatograph and separator, and the selection of a derivative giving an ion which carries a large proportion of the total ion current.

Much effort may have to be expended in order to find a useful derivative which has an intense ion at high enough m/e value to avoid serious interference from other substances, but the improvement obtained can sometimes be quite remarkable. In the case of octopamine[28] the trifluoroacetyl derivative gave the spectrum shown in Fig. 6.18(a).

The ion at m/e 328, while in a region of the spectrum where interference is low, carries only 2% of the total ion current. However, the TMS derivative, whose spectrum is shown in Fig. 6.18(b), looks almost like a typical c.i. spectrum, with an ion at m/e 174 which carries 46% of the total ion current. Although this ion is at a part of the spectrum where interference may be suffered from endogenous substances in urine, plasma or CSF, the gain in sensitivity of a factor of 20 or so more than outweighs the problem of interference, especially as a liquid chromatography technique was used to isolate octopamine and its internal standard from the biological fluids.

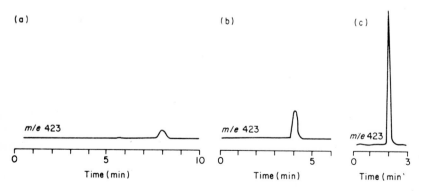

Fig. 6.19.   The improvement in signal height obtained for m/e 423 from prostaglandin $F_{2\alpha}$ methyl ester TMS derivative by increasing the temperature of the SE 30 column: (a) 230 °C; (b) 240 °C; (c) 250 °C.

In searching for a derivative of this nature, the question of retention time must also be taken into account. Ultimately the sensitivity of the analysis depends upon the maximum sample flow into the mass spectrometer source. The signal height obtained will be inversely proportional to the retention time on the column, so that derivatives should be chosen to have as short a retention time as possible, or the temperature of the column can be raised to reduce the retention time. Although a certain derivative may have an ion, say, four times as intense as an ion in a different derivative, it will be inferior from a sensitivity point of view if its retention time is five or six times as great as the same column temperature.

Sometimes, as in the case of the TMS derivative of prostaglandin $F_{2\alpha}$ methyl ester, raising the column temperature to decrease the retention time results in a greater improvement than would be expected. Other things being equal, the peak heights increase because the peaks become narrower, but their area should remain constant. As shown in Fig. 6.19, where the fragment ion at m/e 423 is being monitored, the signal height for a 100 pg injection increases by a factor of 10 or so by raising the column temperature from 230 °C to 250 °C, even though the retention time is reduced by only a factor of about three. The greater improvement could be due to two factors. First, column absorption decreases at higher temperatures, which will result in an improvement for low sample levels, although at the high nanogram level, where absorption is only a small percentage of the amount of sample injected, the improvement will be negligible. Second, fragmentation is increased at higher temperatures, as discussed in Chapter 3, so

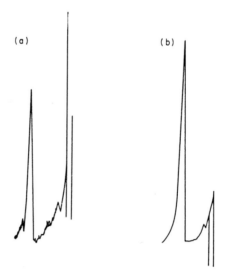

Fig. 6.20. The response at m/e 423 from a 1 pg injection of prostaglandin $F_{2\alpha}$ methyl ester TMS derivative. (a) was obtained six months after (b) on a column which was only used for the analysis of prostaglandin $F_{2\alpha}$. (Reproduced with permission from Ref. 54.)

that the intensity of a fragment ion such as m/e 423 will increase at higher column temperatures. This second effect will of course be counter-productive in those cases where molecular ions are monitored.

Obviously a limit to the reduction in retention time is imposed by interference from the solvent front, which may extend some minutes from the injection point where mixed solvents or silylating agents are used. One way of overcoming this may be to use a much higher loading (e.g. 10 %) of stationary phase for the first few inches of the column.[53] This has the effect of narrowing the solvent front, leading to a better separation of components with a short retention time.

Without a doubt, the most serious limitation to high sensitivity is imposed by adsorption or absorption in the column. In really bad cases it is possible to inject tens of nanograms and obtain a barely detectable signal. In this laboratory, in all cases where a column has been dedicated to the analysis of one particular compound, a steady improvement in sensitivity to that compound has resulted. The importance of this cannot be emphasized too strongly and is underlined by a comparison of the two selected ion profiles in Fig. 6.20. This shows the analysis of the TMS derivative of prostaglandin $F_{2\alpha}$ methyl ester on a dedicated column, with the responses for a 1 pg injection compared over an interval of several months.[54]

To achieve the low femtogram level of detection now being reported for some compounds, it is necessary to saturate the column with nanogram injections of the compound being determined. In the case of the TMS derivative of octo-pamine, a newly conditioned column was just capable of producing a detectable response with an injection of 20 pg,[28] as shown in Fig. 6.21(a). Following 10

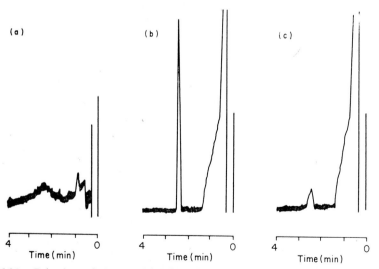

Fig. 6.21. Injection of the tetrakis-TMS derivative of octopamine: (a) 20 pg on a newly conditioned OV 17 column; (b) 10 pg following saturation with 10 injections of 10 ng; (c) a blank injection following the saturation with the 10 ng injections.

injections of 10 ng of the derivative, an injection of 10 pg of the derivative in the silylating agent as solvent gave the response shown in Fig. 6.21(b).

By the end of two days of analysis of 10 pg levels of octopamine, the sensitivity had improved to a point where 1 pg gave a response with a S/N ratio of about 6:1.

The adsorption sites on columns are remarkably specific, because in the case of octopamine the sites distinguish between the unlabelled compound and its $[^2H_3]$ analogue. Even though the column, after saturation with unlabelled compound, was able to give a good response for 10 pg, as shown in Fig. 6.21(b), the

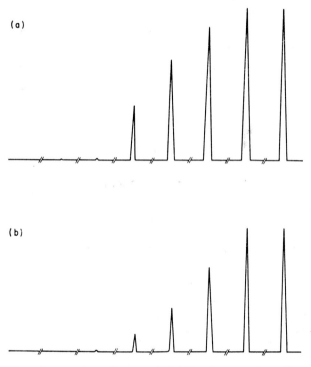

Fig. 6.22. Effect of saturation of a new OV 101 column and an illustration of the specificity of adsorption sites: (a) eight successive injections of 5 ng methyl arachidonate, monitoring the molecular ion at m/e 318; (b) following the injections in (a), the responses for eight injections of 5 ng of $[^2H_8]$arachidonate, monitoring the molecular ion at m/e 326.

detection limit for the labelled compound at that point was still of the order of 20–30 pg. This level did not improve until the column was saturated with 10 ng amounts of labelled compound. This specificity explains the absence of a carrier effect when mixtures of unlabelled and labelled compounds are injected. The same high specificity has been found with the compounds in Fig. 6.13 shown to lack a carrier effect.

The effect of saturating a column, and the specificity of unlabelled versus labelled compound is illustrated in Fig. 6.22. On a newly conditioned 1 % OV 101 column, eight successive injections of 5 ng of methyl arachidonate (21) gave the responses shown in Fig. 6.22(a) for m/e 318, the molecular ion. The first two injections gave no detectable response, but as the syringe, column, separator and source gradually became saturated with the compound, the responses increased, eventually reaching a steady level.

The mass spectrometer was then tuned to m/e 326, the molecular ion of the [^{2}H$_8$] analogue (22), and the same series of eight 5 ng injections repeated. If

21   22

saturation with unlabelled arachidonate had also saturated the adsorption sites for the labelled compound, a response at m/e 326 would have been observable for the first injection. However, as Fig. 6.22(b) shows, this was not the case, and the column had to become saturated with the labelled compound before a steady state was reached. It might be argued that the loss in sensitivity for the first injections of 22 was due to bleed-off of the arachidonate (21) adsorbed on the column during the period necessary to retune the mass spectrometer, but a final injection of unlabelled 21 gave a response similar to the last response in the series of injections shown in Fig. 6.22(a).

Thus, the adsorption sites in the system are able to differentiate between labelled and unlabelled arachidonate, which is consistent with there not being a carrier effect by a large excess of the labelled arachidonate as shown in Fig. 6.13.

As with other compounds, saturation of the column, although leading to an immediate improvement of sensitivity, does not lead to responses of sub-picogram injections until the column has been used for some time at the lower sample levels. Following saturation of the OV 101 column, it was possible to see about 5 pg of methyl arachidonate with a S/N ratio of about 10:1. By the following day sensitivity had improved gradually until an injection of 1 pg and an injection of 100 fg gave the responses shown in Fig. 6.23(a and b). This level of sensitivity was maintained as long as injections were carried out at frequent intervals. If the column was left running for an hour or so with no injections, it was necessary to inject a few 10 pg amounts in order to restore sensitivity to the 100 fg level.

The above comments are not intended to suggest that femtogram amounts of any compound can be detected by single ion monitoring, because the reasons why some compounds are more strongly adsorbed or degraded during their passage through the g.c. column than others are not understood at present. The

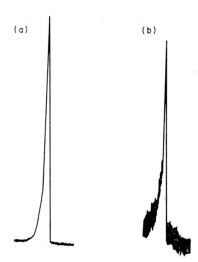

Fig. 6.23.   Injections of methyl arachidonate on the saturated OV 101 column after a day of analyses at the 10 pg level, monitoring *m/e* 318; (a) 1 pg injected; (b) 100 fg injected.

comments are intended to suggest ways in which sensitivity can be improved, and that attention should be focused primarily on the gas chromatograph and molecular separator rather than on the mass spectrometer. Once this is done, without a doubt a great many compounds of biological interest especially will become routinely measurable at the femtogram level, and mass spectrometry will become firmly established as the most sensitive and specific method for the determination of organic compounds.

REFERENCES

1. D. Henneberg, *Z. Anal. Chem.* **170**, 365 (1959).
2. D. Henneberg, *Z. Anal. Chem.* **183**, 12 (1961).
3. G. Schomberg and D. Henneberg, *Chromatographia* **1**, 23 (1968).
4. C. C. Sweeley, W. H. Elliot, I. Fries and R. Ryhage, *Anal. Chem.* **38**, 1549 (1966).
5. R. A. Hites and K. Biemann, *Anal. Chem.* **42**, 855 (1970).
6. J. T. Watson, F. C. Faalkner and B. J. Sweetman, *Biomed. Mass Spectrom.* **1**, 156 (1974).
7. A. E. Banner, *J. Sci. Instrum.* **43**, 138 (1966).
8. A. E. Banner, American Society for Mass Spectrometry, 13th Annual Conference on Mass Spectrometry and Allied Topics, St Louis, Missouri (1965). Proceedings, p. 193.
9. B. W. Wilson, W. Snedden, K. T. Tham and D. A. Willoughby, *Future Trends in Inflammation II. Agents and Actions*, Vol. 6 (J. P. Giroud, D. A. Willoughby, G. P. Velo, editors), Birkhauser-Verlag, Basel, 1976, p. 176.
10. J. Eyem, in *Advances in Mass Spectrometry*, Vol. 7, Heyden, London, 1977, p. 1534.
11. D. E. Green and R. H. Hertel, Symposium on Analytical Advances in Clinical and Medicinal Chemistry, Chicago (1973).

12. W. E. Reynolds, in *Biochemical Applications of Mass Spectrometry* (G. R. Waller, editor), Wiley, New York, 1972, p. 109.
13. D. S. Millington, M. E. Buoy, G. Brooks, M. E. Harper and K. Griffiths, *Biomed. Mass Spectrom.* **2**, 219 (1975).
14. M. G. Horning, W. G. Stillwell, J. Nowlin, K. Lertratanangkoon, D. Carroll, I. Dzidic, R. N. Stillwell and E. C. Horning, *J. Chromatogr.* **91**, 413 (1974).
15. W. F. Bayne, F. T. Tao and N. Crisologo, *J. Pharm. Sci.* **64**, 288 (1975).
16. T. E. Gaffney, C.-G. Hammar, B. Holmstedt and R. E. McMahon, *Anal. Chem.* **43**, 307 (1971).
17. A. K. Cho, B. J, Hodshon, B. Lindeke and G. T. Miwa, *J. Pharm. Sci.* **62,** 1491 (1973).
18. J. T. Biggs, W. H. Holland, S. Chang, P. P. Hipps and W. R. Sherman, *J. Pharm. Sci.* **65**, 261 (1976).
19. W. O. R. Ebbighausen, J. H. Mowat, H. Stearns and P. Vestergaard, *Biomed. Mass Spectrom.* **1**, 305 (1974).
20. I. Jardine, C. Fenselau, M. Appler, M. N. Kan, R. B. Brundrett and M. Colvin, *Cancer Res.* **37** (1977).
21. J. T. Rosenfeld, B. Bowins, J. Roberts, J. Perkins and A. S. Macpherson, *Anal. Chem.* **46**, 2232 (1974).
22. L. Bertilsson, A. J. Atkinson Jr, J. R. Altheus, A. Hartast, J.-E. Lindgren and B. Holmstedt, *Anal. Chem.* **44**, 1434 (1972).
23. K. Green, E. Granstrom, B. Samuelsson and U. Axen, *Anal. Biochem.* **54**, 434 (1973).
24. M. Hamberg, *Anal. Biochem.* **55**, 368 (1973).
25. F. Karoum, J. C. Gillin, R. J. Wyatt and E. Costa, *Biomed. Mass Spectrom.* **2**, 183 (1975).
26. B. Samuelsson, M. Hamberg and C. C. Sweeley, *Anal. Biochem.* **38**, 301 (1970).
27. H. T. Cory, P. T. Lascelles, B. J. Millard, W. Snedden and B. W. Wilson, *Biomed. Mass Spectrom.* **3**, 117 (1976).
28. B. J. Millard, P. A. Tippett, M. W. Couch and C. M. Williams, *Biomed. Mass Spectrom.* **5** (1) (1978).
29. W. D. Reynolds and R. Delongchamp, *Biomed. Mass Spectrom.* **2**, 276 (1975).
30. B. J. Millard and M. G. Lee, *Biomed. Mass Spectrom.* **2**, 78 (1975).
31. B. J. Millard, *Chem. Ind. (London)* 388 (1976).
32. G. H. Draffan, R. A. Clare and F. M. Williams, *J. Chromatogr.* **75**, 45 (1973).
33. J. M. Strong and A. J. Atkinson, *Anal. Chem.* **44**, 2287 (1972).
34. J. N. T. Gilbert and J. W. Powell, *Biomed. Mass Spectrom.* **3**, 1 (1976).
35. S. Agurell, L. O. Boreus, E. Gordon, J.-E. Lindgren, M. Ehrnebo and U. Lonroth, *J. Pharm. Pharmacol.* **26**, 1 (1974).
36. T. Walle, J. Morrison, K. Walle and E. Conradi, *J. Chromatog.* **114**, 351 (1975).
37. L. Bertilsson and L. Palmer, *Science* **177**, 74 (1972).
38. A. Frigerio, G. Belvedere, F. De Nadai, C. Fanelli, C. Pantarotto, F. Riva and C. Morselli, *J. Chromatogr.* **74**, 201 (1972).
39. R. Fanelli and A. Frigerio, *J. Chromatogr.* **93**, 441 (1974).
40. H. R. Sullivan, F. J. Marshall, R. E. McMahon, E. Anggard, L. M. Gunne and J. H. Holmstrand, *Biomed. Mass Spectrom.* **2**, 192 (1975).
41. R. L. Foltz, M. W. Couch, M. Greer, K. N. Scott and C. M. Williams, *Biochem. Med.* **6**, 294 (1972).
42. A. G. Smith and C. J. W. Brooks, *Biomed. Mass Spectrom.* **4**, 258 (1977).
43. S. B. Matin, J. H. Karam, P. H. Forsham and J. B. Knight, *Biomed. Mass Spectrom.* **1**, 320 (1974).
44. P. A. Clarke and R. L. Foltz, *Clin. Chem.* **20**, 465 (1974).

45. D. C. K. Lin, A. F. Fentiman Jr, R. L. Foltz, R. D. Forney and I. Sunshine, *Biomed. Mass Spectrom.* **2**, 206 (1975).
46. M. Claeys, G. Muscettola and S. P. Markey, *Biomed. Mass Spectrom.* **3**, 110 (1976).
47. G. C. Ford, N. J. Haskins, R. F. Palmer, M. J. Tidd and P. H. Buckley, *Biomed. Mass Spectrom.* **3**, 45 (1976).
48. S. B. Matin, S. H. Wan and J. B. Knight, *Biomed. Mass Spectrom.* **4**, 118 (1977).
49. R. J. Weinkam, M. Rowland and P. Meffin, *Biomed. Mass Spectrom.* **4**, 42 (1977).
50. N. Neuner-Jehle, F. Etzweiler and G. Larske, *Chromatographia* **6**, 211 (1973).
51. K. Grob and H. Jaeggi, *Anal. Chem.* **45**, 1788 (1973).
52. J. G. Leferink and P. A. Leclerq, *J. Chromatogr.* **91**, 385 (1974).
53. W. Snedden and B. W. Wilson, private communication (1976).
54. B. W. Wilson and W. Snedden, Varian MAT User's Meeting, Cardiff (1975).

SOME USEFUL STATISTICS EQUATIONS AND TABLES

VARIANCE AND STANDARD DEVIATION

The variance v of a set of results x_1 to x_n is given by the sums of the squares of the differences between each result and the mean, divided by the number of results minus one.

Thus,

$$v = \frac{\sum_{i=1}^{i=n}(x_1 - \bar{x})}{(n-1)}$$

where there are n results, and $\bar{x}$ is the mean of the n results.

The calculation can be made simpler by the transformation

$$\sum_{i=1}^{i=n}(x_i - \bar{x}) = \sum_{i=1}^{i=n}(x_i^2) - \frac{\left(\sum_{i=1}^{i=n}x_i\right)^2}{n}$$

The standard deviation $s = \sqrt{v}$.

COEFFICIENT OF VARIATION

The coefficient of variation (c.v.) is the ratio of the standard deviation to the mean, expressed as a percentage:

$$\text{c.v.} = 100\frac{s}{\bar{x}}\%$$

STANDARD ERROR OF THE MEAN

From the limited number of results n, the mean can be taken and the standard deviation can be calculated as above. However, the object is to determine the true result of the particular experiment. The best estimate of the standard deviation of $\bar{x}$ about the true result is given by the standard error of the mean (s.e.m.)

$$\text{s.e.m.} = \frac{s}{\sqrt{n}}$$

This takes account of the number n of results used in the calculation of the mean, and shows that quadrupling the number of results will reduce the standard error by a factor of 2.

LIMITS OF ERROR OF THE MEAN

The limits of error give the range in which the true result will lie for a given probability level (e.g. 90 or 95%):

$$\text{limits of error} = \pm t \times \text{standard error}$$

where t is Student's t given in the next table.

STUDENT'S t TEST

This is used to compare two samples of values of a quantity. The test decides whether the difference between two sample means is significant.

$$t = \frac{|m_A - m_B|}{s\sqrt{(1/n_A + 1/n_B)}}$$

where m_A, m_B are the means of the two sets of results, s is the overall standard deviation of the combined sets of results, and n_A and n_B are the numbers of results in each set A and B.

If the calculated value of t is greater than the value given in the table below, the difference in the means is considered to be significant. The tabulated values are for a probability of 95%, i.e. there is a 95% probability that the means are significantly different if t exceeds the value in the table. The degrees of freedom are $n_A + n_B - 2$.

Table of values of t at the 0.95 probability level

Degrees of freedom	1	2	3	4	5	6	7	8	9
t	12.71	4.30	3.18	2.78	2.57	2.45	2.37	2.31	2.26
Degrees of freedom	10	15	20	25	30	40	60	120	∞
t	2.23	2.13	2.09	2.06	2.04	2.02	2.0	1.98	1.96

VARIANCE RATIO TEST

This is another method of comparing two sample values of a quantity. The variances of each set of results are compared to decide whether the two sets of results come from the same population. The ratio F of the larger variance divided by the smaller variance is compared with the value of F from the table to see whether the difference between the variances is significant. If the calculated value is greater than that in the table, there is a 95% probability that the variances are significantly different.

Values of F at the 0.95 probability level

Degrees of freedom of the lesser variance	Degrees of freedom of the greater variance							
	1	2	3	4	6	8	10	∞
1	161	199	216	225	239	237	242	254
2	18.5	19	19.2	19.2	19.3	19.4	19.4	19.5
3	10.1	9.55	9.28	9.12	8.94	8.85	8.79	8.53
4	7.71	6.94	6.59	6.39	6.16	6.04	5.96	5.63
6	5.99	5.14	4.76	4.53	4.28	4.15	4.06	3.67
8	5.32	4.46	4.07	3.84	3.58	3.44	3.35	2.93
10	4.96	4.10	3.71	3.48	3.22	3.07	2.98	2.54
∞	3.84	3.00	2.60	2.37	2.10	1.94	1.83	1.00

LINEAR REGRESSION

The equation of the regression line y/x which minimizes the sums of the squares of the vertical (y) deviations from the straight line is given by

$$y - \bar{y} = b(x - \bar{x})$$

where $\bar{y}$, $\bar{x}$ are the means of all the values of y and x respectively, and b is the slope, given by

$$b = \frac{\sum_{i=1}^{i=n}(x_i y_i) - \sum_{i=1}^{i=n}(x_i)\sum_{i=1}^{i=n}(y_i)/n}{\sum_{i=1}^{i=n}(x_i^2) - \sum_{i=1}^{i=n}{}^2(x_i)/n}$$

where n is the number of points used to carry out the regression.

If the line is required to pass through the point (x_0, y_0) (frequently the origin), the slope of the line is given by

$$b = \frac{\sum_{i=1}^{i=n}(x_i - x_0)(y_i - y_0)}{\sum_{i=1}^{i=n}(x_i - x_0)^2}$$

The above regressions assume that the variance in y is greater than the variance in x. If it is shown before carrying out the regression that the variance in y is less than that in x, the above equations can be used by transposing x and y in them.

VARIANCE OF A VALUE OF x ESTIMATED FROM A SINGLE VALUE OF y AND THE REGRESSION LINE

The variance v_x of this value of x is given by

$$v_x = \frac{v_y}{b^2}\left[1 + \frac{1}{n} + \frac{(y - \bar{y})^2}{b^2 \left(\sum_{i=1}^{i=n}(x_i - x)^2\right)}\right]$$

where v_y is the variance of the values of y used in the regression calculation and b is the slope of the regression line.

If m repeats are made to determine the value of y, the equation becomes

$$v_x = \frac{v_y}{b^2}\left[\frac{1}{m} + \frac{1}{n} + \frac{(y - \bar{y})^2}{b^2 \left(\sum_{i=1}^{i=n}(x_i - x)^2\right)}\right]$$

CORRELATION COEFFICIENT

A regression line is meaningful only if there is a significant linear correlation between the sets of x and y values. This can be tested by means of the correlation coefficient r, calculated as follows:

$$r = \frac{\sum_{i=1}^{i=n} x_i y_i - \sum_{i=1}^{i=n}(x_i)\sum_{i=1}^{i=n}(y_i)/n}{\sqrt{\left[\left\{\sum_{i=1}^{i=n}(x_i^2) - \sum_{i=1}^{i=n}{}^2(x_i)/n\right\}\left\{\sum_{i=1}^{i=n}(y_i^2) - \sum_{i=1}^{i=n}{}^2(y_i)/n\right\}\right]}}$$

where the terms have the usual significance.

If the value of r calculated as above is greater than the value of r in the following table (at the 0.95 probability level), then there is a 95% probability that there is a significant correlation between x and y, and a regression line can be calculated. The number of degrees of freedom in the table is given by $(n - 2)$ where n is the number of pairs of results x, y.

Theoretical values of the correlation coefficient

Degrees of freedom	2	3	5	7	10	30	100
Correlation coefficient	0.95	0.88	0.75	0.67	0.58	0.35	0.20

RECOMMENDED READING

M. J. Moroney, *Facts from Figures*, 2nd Edn, Penguin Books, Harmondsworth, 1953.
L. Saunders and R. Fleming, *Mathematics and Statistics for Use in the Biological and Pharmaceutical Sciences*, 2nd Edn, The Pharmaceutical Press, London, 1971.
D. Colquhoun, *Lectures in Biostatistics*, Clarendon Press, Oxford, 1971.
C. G. Paradine and B. H. P. Rivett, *Statistical Methods for Technologists*, English University Press, London, 1968.
R. A. Fisher and F. Yates, *Statistical Tables for Biological, Agricultural and Medical Research*, Longman, Harlow, 1974.

SUBJECT INDEX

A

Accelerating voltage alternator, 27
Accelerating voltage stability, 27
Accuracy, 83
    in construction of calibration lines, 86–89
Accurate mass determination, 13–14
Allopurinol, 110
Ammonia, as c.i. reagent gas, 7
Analogue to digital converter, 29, 41
    errors caused by, 42–43
Androstane diols, single ion monitoring of, 134
Appearance potential, 2
Arachidonate, labelled, 156
    lack of carrier effect with, 140
    saturation of columns with, 156–158

B

Background, subtraction of, 20
Bandwidth, effect on scanning, 119
Biemann library search system, 36
Binary mixtures, response ratios for, 95–96
Bleed absorbing column, 20
Brunée separator, 20

C

Calibration lines, 56, 73–89

change in slope of, 84–85
for octopamine, 83
for phenylalanine in blood, 112
for prostaglandin $F_{2\alpha}$, 81–82
for putrescine, 105
for p-tyramine, 107
Californium 252 ionization, 11
Capillary columns, 151
Carrier effect, lack of, 81, 139–142
Centroid, 30–31, 42
Charge exchange, 7–8
    spectrum of heroin, 8
Chemical ionization, 5–7, 149–151
    for amino acid determination, 111
    for tolbutamide determination, 114
    spectrum of primidone metabolite, 150
        of quinidine, 111
        of tetrahydrocortisone, 10
Chlorpromazine, 4
Clerc library search system, 37
Coefficient of variation, 53, 57, 80, 97, 146
    at various dwell times, 126
Cold spots, 22
Column bleed, 20
Computers, 28–37
    control of mass spectrometers, 31–33, 126–127
Correlation coefficient, 74

D

Daly collector, 17